A Miscellany of Bats

A Miscellany of Bats

M. Brock Fenton and Jens Rydell

PELAGIC PUBLISHING

Published by Pelagic Publishing
20–22 Wenlock Road
London N1 7GU, UK

www.pelagicpublishing.com

A CIP record for this book is available from the British Library

ISBN 978-1-78427-294-4 Paperback
ISBN 978-1-78427-295-1 ePub
ISBN 978-1-78427-296-8 ePDF

https://doi.org/10.53061/LLFU5654

Typeset in Palatino Linotype and Myriad Pro by Trevor Johnson

Printed in England by Swallowtail

Front cover images, clockwise from top left: Wahlberg's Epauletted Fruit Bat, Hildegard's Tomb Bat (*JR*), Daubenton's Bat (*JR*), Yellow-winged Bat (*JR*), Hoary Bat, Common Sheath-tailed Bat (*JR*)
Back cover images: Jamaican Fruit Bat, Common Vampire Bat
Title pages: Pale Spear-nosed Bat
Opposite: Heller's Broad-nosed Bat

Contents

Jens Rydell at an entrance to the Hemsö Fortress in Sweden, a relict of the Cold War. Built to withstand a direct nuclear hit, part of this fortress is now used by hibernating Northern Bats. Jens and Johan were looking for sites used by swarming bats (see page 166). *Photo by Johan Eklöf, 30 August 2017*

Preface: Bats and Jens

There are more than 1,400 different species of bats worldwide. These are classified into 21 families (see table on page xi). Here a family is a group of close relatives (in an evolutionary sense). Although we usually think of them as flying in the air, bats are land mammals, occurring everywhere except in Antarctica, the high Arctic and some remote oceanic islands.

The purpose of this book is to introduce you to the extraordinary and diverse world of bats: what they are, what they do and what we know about them. You will see that bats connect with people by way of various myths and superstitions, as well as through the science of studying them. While bats' alleged role in the COVID-19 pandemic has not endeared them to people in general, it has highlighted the possible importance of these animals because of their astonishing capacity for neutralizing the negative impacts of some viruses.

Wings and flight are of course the two main distinctive characteristics of bats, making them easy to recognize anywhere. Bats have many other features – some basically mammalian, others relating to echolocation – but none of them is exclusive to, or characteristic of, bats. At least some species of bats have long lifespans, but their reproductive output is low, typically just one litter a year. Most species bear one young in a litter but some have twins.

Bats are not blind and do not get tangled in people's hair. Some eat astonishing numbers of insects, and others pollinate flowers. Populations of bats may be limited by the quantity of roost spaces available, while some species readily roost in human structures from buildings to mines.

Unfortunately, suddenly and unexpectedly, co-author Jens Rydell (1953–2021) passed away just when this book was about to be finished. Johan Eklöf, one of his many former students, reflects on what Jens might have said here:

He would have kept it short, letting the reader quickly move on to the main purpose at hand: the photographs and stories behind the research. The saying 'a picture is worth a thousand words' is perhaps an old cliché, but still very true, and it was Jens' motto. He photographed things nobody else saw: tiny insects, snowflakes, fossils, the beauty of darkness and of course, bats. Jens was a storyteller, and every single photo had its own narrative. He never beautified his photos. If a bat was flying outside a public restroom or in front of graffiti-sprayed walls, that was an important part of the picture too, and this often encapsulates the real beauty in his work. Jens rarely called himself a nature photographer. The camera was rather a component in his scientific toolbox, just as important as the bat detector and the little black notebook he always carried around. His bookshelves contained close to a hundred such notebooks, all the same design, ranging in date from the late 1970s to March 2021; these represent a goldmine for future generations of researchers. The same goes for the hard drives with thousands of photos, many detailed enough to study the bats as if they were in hand. In 2014, Jens published a paper ('Photography as a low-impact method to survey bats') on how this approach could be not only a way to study bats but perhaps even the best tool for the job. A few years ago, I suggested to Jens that he should sum up all his travels, photographic expeditions, and scientific journeys in a book, like an old-school adventurer sharing his discoveries. There were just so many tales to be told. He simply shook his head; he did not want to write about himself. But this is a book about a bat...

Some of you (hopefully reading this) are mentioned in this books' acknowledgements. But without Jens' presence, many names are still missing, and unfortunately the book about all of you was never realized. Instead, this is Jens' last publication – and in a way a legacy and a summary of his contribution to bat science. Jens was proud of it, really looking forward to finally sharing the writing of a popular science book with Brock, his old friend and colleague with whom he published many scientific papers. But in Jens' eyes, this was not a summary of a career, it was just a step on the way. He was still full of new ideas. The very same day he passed away, I had talked to him on the phone, and he said: 'Johan, now I know what we need to do next summer...'. It concerned Northern Bats, the species he had studied his whole life, and which had comprised a major part of his doctoral study in the 1980s. He was ever thus: tireless in his curiosity about and enjoyment of bats.

Nomenclature

All species have scientific names, to increase precision of communication among scientists. Bats are no exception. Scientific names are known as binomials, hence comprising two parts: the genus name and the epithet. By convention, scientific names are always italicized. In this book we use common names in the main text and captions, relying on the Appendix to link common and scientific names.

Now we consider the names of bats in particular. Roberts's Flat-headed Bat from southern Africa is a good place to start. Roberts was the person who first described the species, which has a flattened skull. The scientific name is *Sauromys petrophilus,* which roughly translates to 'lizard-loving rock mouse'.

Common names in English are certainly easier to get a handle on for the beginner, but the lack of standardization of bats' common names can be confusing. A bat known as the Little Brown Myotis (*Myotis lucifugus*) is called a 'Little Brown Bat' in some cases. Unlike birds, for instance, many bats do not have standardized common names.

Scientific names place species in a classification reflecting the organism's evolutionary history. *Sauromys petrophilus* belongs to the family Molossidae (free-tailed bats) and *Eptesicus fuscus* belongs to the family Vespertilionidae (vesper bats). Families are made up of genera (plural of genus), genera comprise groups of species. The number of species in a genus can vary a good deal: *Sauromys* has one species, but *Eptesicus* has about 25. The families of bats (common names and scientific names) are listed in the Appendix.

After the discovery of blood-feeding bats, many bats were given scientific names with a 'vampire' root (e.g. *Vampyrum, Vampyressa, Vampyrops*), and others were called 'false vampire bats'. The family of bats known as 'false vampire bats' were thought to be like 'real' vampire bats and eat the blood of other animals. To date there is no evidence of this, but the name lingers on. To make matters worse, scientific names often change to reflect new ideas about boundaries between species and changes in classification. We encountered a 'good' example when preparing this manuscript: species in the family Miniopteridae are known as 'bent-winged bats', yet the common names of the species we researched all were 'long-fingered bats'.

Common name	Scientific name	No. species	Distribution
Yinpterochiroptera *includes 7 families*			
Old World fruit bats	Pteropodidae	~200	Africa, Asia, Australia and some Pacific Islands
mouse-tailed bats	Rhinopomatidae	~6	Africa, southern Asia
bumblebee bats	Craseonycteridae	1	Southeast Asia
horseshoe bats	Rhinolophidae	~97	Eurasia, Africa, Australia
Old World leaf-nosed bats	Hipposideridae	~90	Africa, Southeast Asia, Australia
trident bats	Rhinonycteridae	~10	Africa, Southeast Asia
false vampire bats	Megadermatidae	~5	Africa, Southeast Asia, Australia
Yangochiroptera *includes 14 families*			
slit-faced bats	Nycteridae	~16	Africa, Southeast Asia
sheath-tailed bats	Emballonuridae	~54	Pantropical, Americas, Africa, Southeast Asia, Australia
New World leaf-nosed bats	Phyllostomidae	~220	Americas, Caribbean Islands
moustached bats	Mormoopidae	~12	Americas, Caribbean Islands
bulldog bats	Noctilionidae	~2	Americas, Caribbean Islands
smoky bats	Furipteridae	~2	Tropical Americas
New World disk-winged bats	Thyropteridae	~5	Tropical Americas
Old World disk-winged bats	Myzopodidae	~2	Madagascar
New Zealand short-tailed bats	Mystacinidae	~2	New Zealand
funnel-eared bats	Natalidae	~12	Tropical Americas
free-tailed bats	Molossidae	~113	Eurasia, Africa, Asia, Australia, Americas
bent-winged bats	Miniopteridae	~29	Eurasia, Africa, Asia, Australia
wing-gland bats	Cistugidae	~2	Southern Africa
vesper bats	Vespertilionidae	~455	Worldwide except Antarctica, Arctic

Bat Classification now. Living bats are placed in 21 families classified in two major groups (suborders) which have lovely, unwieldy names: Yinpterochiroptera and Yangochiroptera.

Acknowledgements

We thank Mark Brigham and Erin Fraser for reading and commenting on the entire manuscript. We thank Darrian Washinger for her work on the sound pictures. We are very grateful to Ernest Seamark, Amanda Stronza (BCI), Al Hicks, Scott Pedersen, Gerald Kerth, and Brian Keeley for permitting us to use their pictures. The exceptional facilities and personnel of Lamanai Outpost Lodge in Belize were fundamental to our many visits there.

In addition to Johann Eklöf, Jens wished to express his gratitude to the following: Damian Milne and Thomas Madsen (Australia), Luisa Rodrigues, Hugo Rebelo, Helena Santos, Pedro Alves, Bruno Silva and Silvia Pereira Barreiro (Portugal), Ernst Herman Solmsen (Costa Rica), Eran Amichai, Arjan Boonman, Ivo Borissov, Ofri Eitan, Yossi Yovel and Carmi Korine (Israel), Tomasz Kokurewicz and Grzegorz Apoznański (Poland), Gunars Petersons, Jurgis Šuba, Alda Stepanova, Viesturs Vintulis and Ilze Brila (Latvia), Raphaël Arlettaz (Switzerland), Chen-Han Chou, Heng-Chia Chang, Hsi-Chi Chen, Hong-Chang Chen, Kuang-Lung Huang and Hsue-Chen Chen (Taiwan), Antonio Guillén-Servant, Anglelica Menchaca and Rodrigo Medellin (Mexico), Javier Juste, Carlos Ibañez, Sonia Sánchez Navarro, Juan Quetglas, Domingo Trujillo and Rubén Barone (Spain), Bert Wiklund, Anita and Lee Hildsgaard Rom (Denmark), Jeroen van der Kooij, Keith Redford, Tore Christian Michaelsen and Kristoffer Böhn (Norway), Jacques Sirgent (France), Matti Masing (Estonia), Marco Riccucci and Danilo Russo (Italy), Gareth Jones and Roger Ransome (England), Sara Bumrungsri, Tuanjit Sritongchuay, Kanuengnit Wayo, and C. E. Nuevo Diego (Thailand), Ravi Umadi and Sumit Dookia (India), Magnus Gelang, Stefan Pettersson, Espen Jensen, Andreas Olsson, Sabine Lind, Hans Fransson and others (Sweden), Simon Musila, Paul Webala, Robert Syingi, Mike Bartonjo, Beryl Makori, Simon Masika and especially Aziza Zuhura (Kenya). Also, many thanks to family, friends and all owners and attendants of houses, castles, mills, barns, cellars, ruins, churches, wells, gardens and many other bat places.

In addition to Sherri Lee Fenton, Brock is very grateful to the people who have enriched the bat world for him: Lalita Acharya, Amanda Adams, Rick Adams, Hugh Aldridge, Scott Altenbach, Doris Audet, Robert Barclay, Dan Becker, Gary Bell, Enrico Bernard, Neil Boyle, Hugh Broders, Beth Clare, Karen A. Campbell, Gerry Carter, David and Meg Cumming, Christina Davy, Betsy Dumont, Neil Duncan, Miranda Dunbar, Yvonne Dzal, Johan Eklöf, Francois Fabianek, Paul Faure, Eleanor Fenton, Ted Fleming, Erin Fraser, Fred Frick, Alan Grinnell, Jon Hayes, John Hermanson, James Fullard, Robert Herd, Brian Hickey, Al Hicks, Ying-Yi Ho, Roy Horst, Melissa Ingala, Dave Johnston, Gareth Jones, Teresa Kearny, Brian Keeley, Susan Koenig, Burton Lim, Lauren MacDonald, Alistair MacKenzie, Gary McCracken, Liam McGuire, Gray Merriam, Derek Morningstar, Cindy Moss, Samira Mubareka, Ulla Norberg, Martin Obrist, Dara Orbach, Teri Orr, Scott Pedersen, Paul Racey, John Ratcliffe, Kobie and Naas Rautenbach, Dan Riskin, Danilo Russo, Uli Schnitzler, Ernest Seamark, Price Sewell, Jim Simmons, Nancy Simmons, Mark Skowronski, Angelo Soto-Centeno, Kelly Speer, Sharon Swartz, Don Thomas, Toby Thorne, Geoff Turner, Merlin Tuttle, Terry Vaughan, Martin Vonhof, Sean Werle, Damion Whyte, and Yossi Yovel.

Our biggest debt is of course to the bats themselves.

1 Introducing bats

A Striped Hairy-nosed Bat from Belize. These little-known 15 g bats are widespread but uncommon. We do not know what they eat or where they roost.

Wings and size

Wings make a flying bat, here a Western Barbastelle, unmistakable. But a glimpse of a flying bat usually provides little indication of its size. Bats are small mammals, adults range in weight from 2 g to about 1500 g but most are less than 50 g. In the United States, Canada and Europe, the largest species weigh about 50 g. At 1,100 g, Giant Golden-crowned Flying Foxes have a wingspan of 1.7 m, putting them among the largest of bats. At 2 g (wingspan 15 cm), Kitti's Hog-nosed Bats are the smallest.

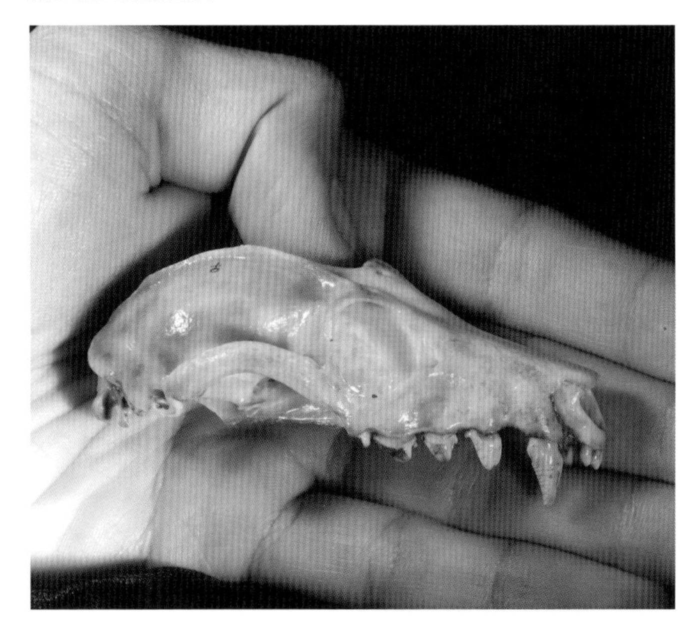

This Western Barbastelle is small to medium in size (11 g, 28 cm wingspan) for a bat of the north temperate zone. This insectivorous species is widespread in Eurasia. Its ears and face are distinctive features for identification. Females bear one or two young a year. They usually hibernate in underground situations in caves or abandoned mines. Western Barbastelles are a hot topic, having been declared Bat Species of the Year for 2020–2021 by Batlife Europe. *JR*

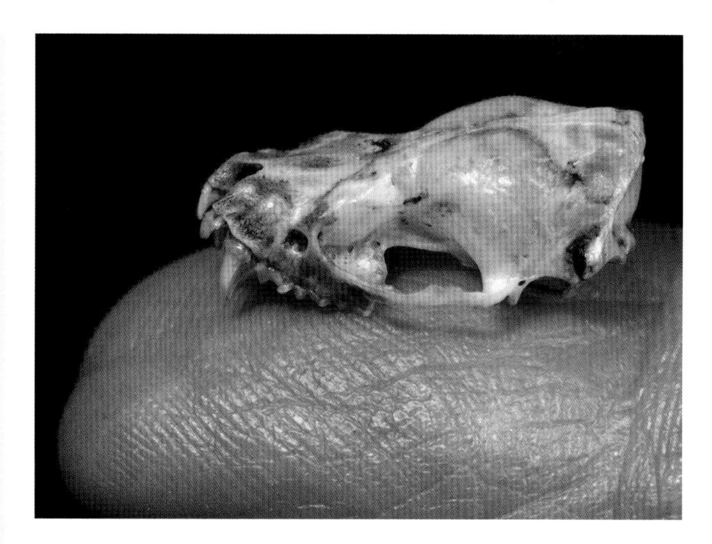

Comparing the skulls of a Giant Golden-crowned Flying Fox (lower left) and a Big Brown Bat (15 g) (right) illustrates the size range of these animals. The Big Brown Bat skull is on MBF's finger. The flying fox occurs in the Philippines. It eats mainly fruit. Females have one young per litter, and one litter per year. They roost by day in trees and are threatened by the bushmeat trade. Big Brown Bats are insectivorous and widespread in North America. They roost in hollows in summer, often in buildings. In winter they hibernate, sometimes in buildings, or in caves or mines. Females have one or two young per litter, and one litter a year.

At 170–200 g, the Spectral Bat is the largest bat species in the New World. This animal-eating species is slightly larger than an Egyptian Fruit Bat (overleaf), which eats mainly fruit.

When the focus is bats, it is helpful to remember three important points. First, these animals are nocturnal. Combined with their generally small size, this can make them difficult to spot, by night or by day in fact because it is easy for them to hide. Second, small size limits the kinds of research possible. Third, bats share a number of basic features with other mammals, including humans: they are warm-blooded, females bear live young and feed their pups milk.

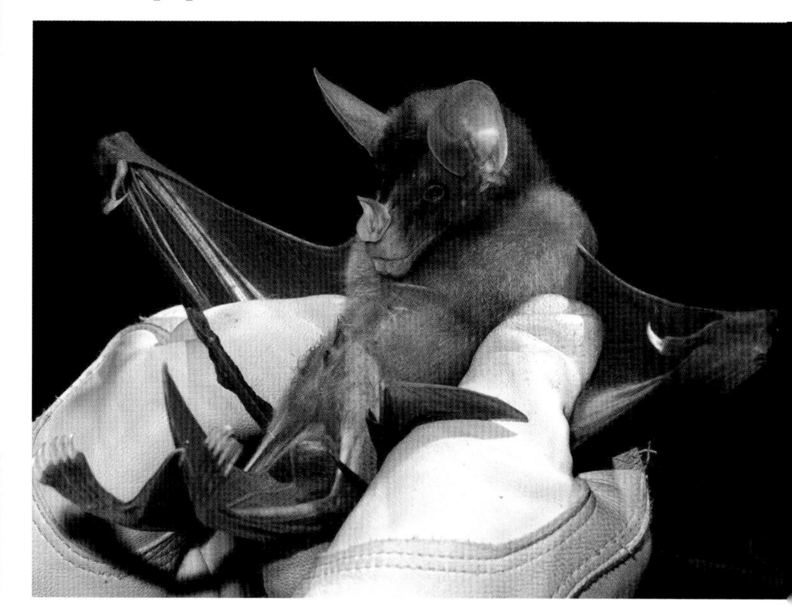

A Spectral Bat from Belize fills two gloved hands.

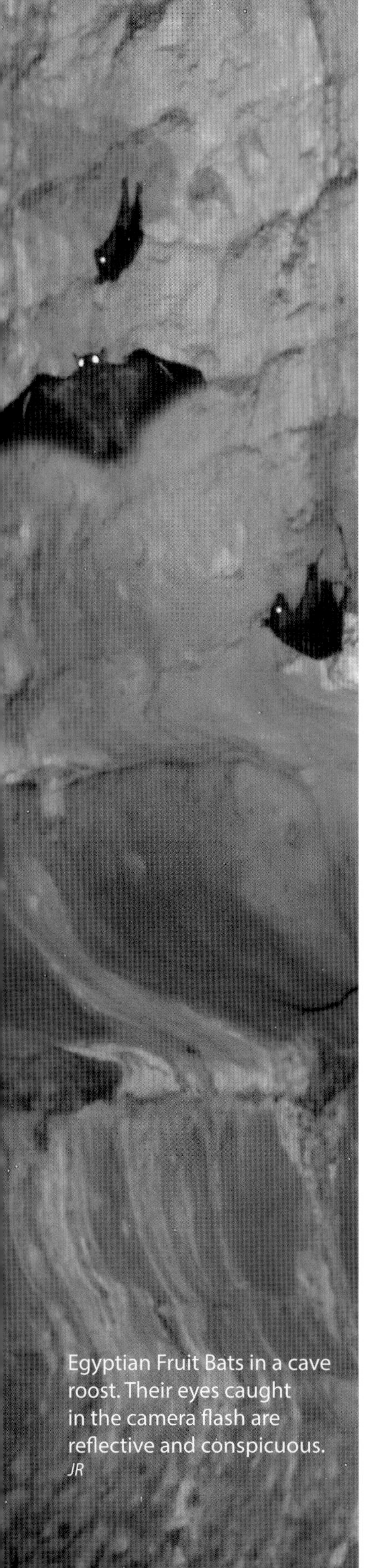
Egyptian Fruit Bats in a cave roost. Their eyes caught in the camera flash are reflective and conspicuous. *JR*

There are still many unanswered questions about this fascinating and diverse group of mammals. We will pose some of these questions in the following pages, giving you a chance to think about possible solutions and perhaps providing inspiration for your research. Throughout this book we link pictures and words to reinforce the stories and facts about bats.

Blind as a bat

Contrary to myths and sayings, bats are not blind. None of the 1,400+ species of bats has lost its sense of vision, and the images of bat faces we include in this book show their eyes very clearly. Why do people think bats are blind? The discovery of echolocation (see Chapter 3) left the impression that bats had biosonar abilities and did not need to see. Furthermore, they are small and many have small eyes. In bright light, a squinting bat may appear to lack eyes altogether. Compared to echolocation, we know much less about the role that vision plays in the lives of bats. Pteropodid bats have very good nocturnal vision, due in part to the presence of a special reflective layer (tapetum lucidum) behind the retina.

Bats' eyes vary in size between species. Some have large, prominent eyes, others much smaller ones. But in a room with windows, a released bat that can echolocate does not fly directly into a window (unlike most birds would do). Echolocation tells the bat the truth about a glass window, that it is not a clear opening. Sometimes, however, when a bat is first released in a room it flies about but then suddenly may fly into a window. In this case the bat appears to switch its sensory attention from echolocation to vision. At a research station in Zimbabwe, this behaviour was particularly obvious in Egyptian Slit-faced Bats which not only flew into window glass but also into glass fronts of cupboards reflecting vegetation outside.[1] In an experiment with Big Brown Bats, researchers showed that these animals used different combinations of auditory and visual cues.[2] Performance in target location improved when these bats used both visual and acoustic cues.[3] This important finding reminds us that bats take advantage of whatever information is available to them. How bats integrate information from vision and from echolocation remains an area of active study.

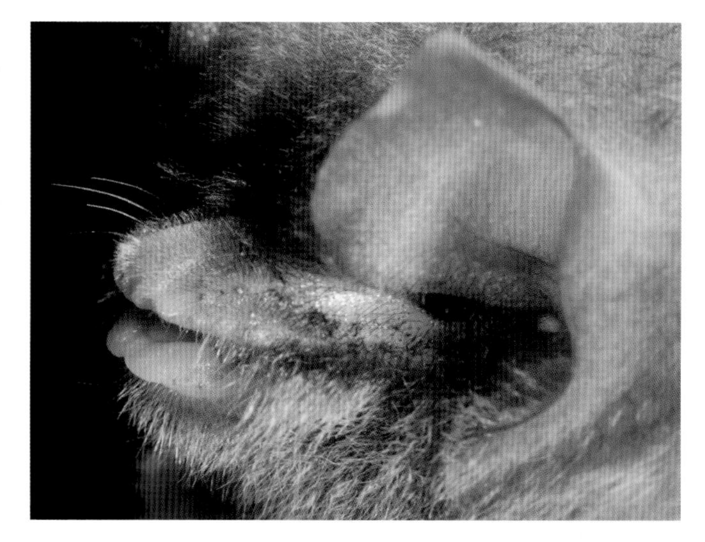

The eyes of a Mexican Funnel-eared Bat (5 g) are partly covered by its ears and nestled in long hairs.

A Little Brown Myotis is slightly larger (8 g) and its eyes are not hidden by fur.

The still larger Meheyli's Horseshoe Bat (14 g) has larger, more prominent eyes. *JR*

Catching and identifying bats

Research into bats often requires catching them, while of course minimizing disturbance and harm. Typically, we use mist nets and harp traps, but this also means deciding where to set the nets and traps and how to protect them and the bats they catch. The disadvantage of mist nets is disentangling each bat without damaging the bat or having it bite you. We think that harp traps are 'better' because the bats do not get tangled, but you still need to remove bats from the trap and they can still bite! The advent of mist nets and bat traps had a strong positive impact on bat research. Bat biologists should take care to minimize the chances of spreading SARS-CoV-2 (which causes COVID-19 – see page 210) to bats, hence the masks.

The harp trap is a 2 m by 2 m frame supporting two vertical rows of fine lines. When set in a bat flyway, bats hit the lines and slide down into a holding bag. The long narrow opening to the bag stops bats flying out, and slippery plastic sheets keep them from climbing out. The trap is set in front of the entrance to a mine used as a hibernation and swarming site. The researchers here are Gwynne Domashinksi, MBF and Lily Hou.

Little Brown Myotis on trap strings

Little Brown Myotis in the bat trap. These traps are very effective for catching bats when set correctly in the 'right' place(s).

The Little Brown Myotis tangled in a mist net will need to be carefully disentangled. Like harp traps, mist nets can be very effective for catching bats when used correctly.

Some salient features of bats

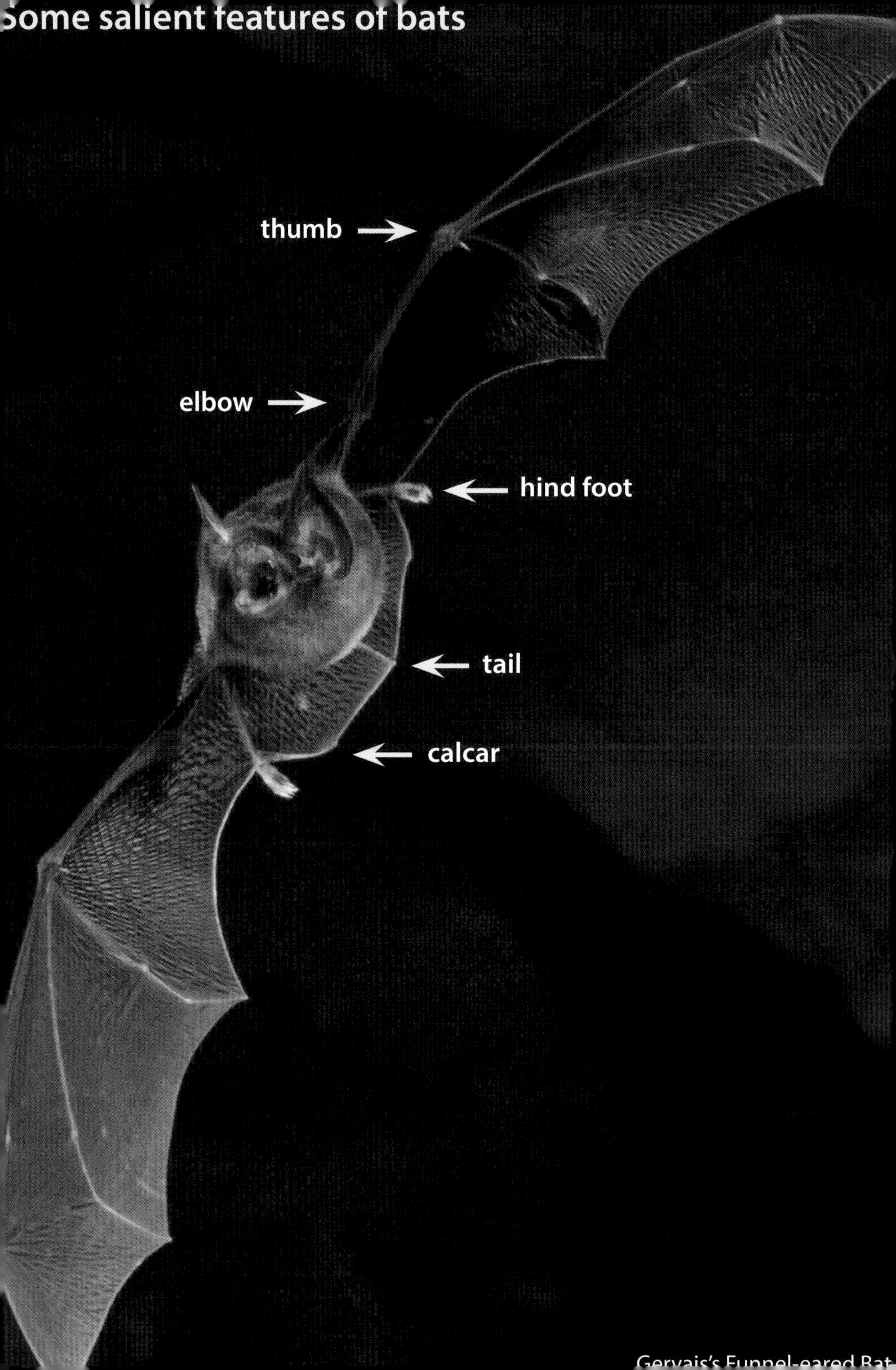

Gervais's Funnel-eared Bat

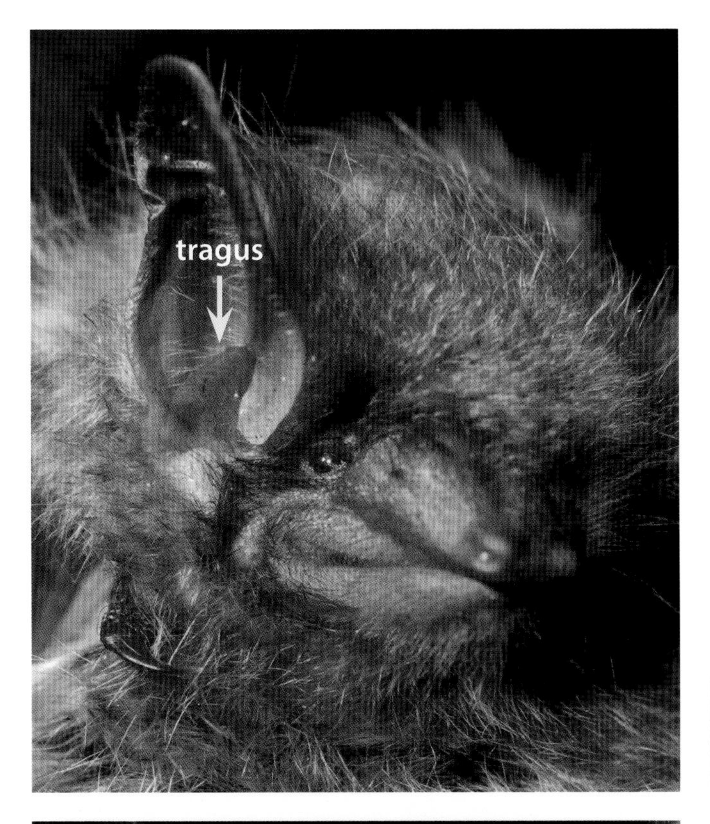

The tragus identified on a Tricolored Bat.

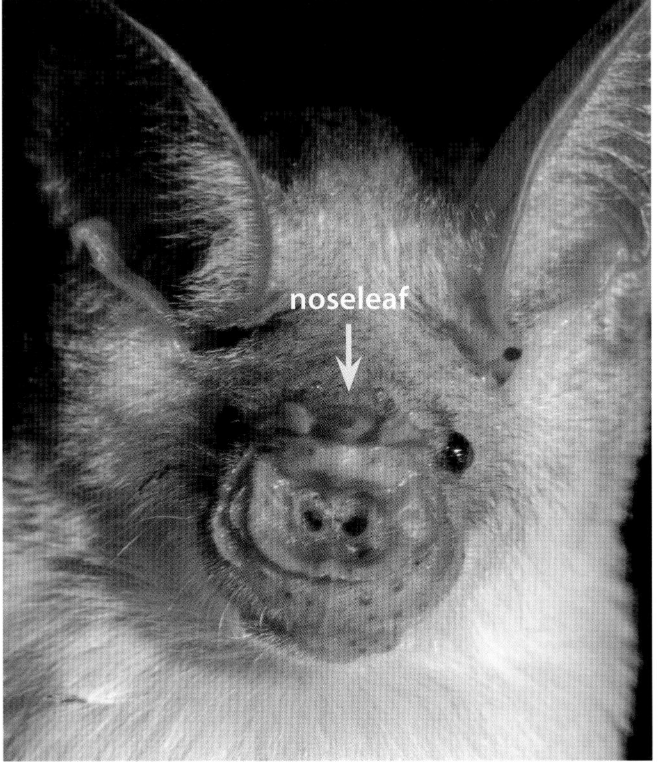

A Trident Leaf-nosed bat.

The insectivorous bats on these pages were photographed in Cuba (GFEB), Quebec, Canada (TB) and in Israel (TLNB). Note differences in eye-size. The GFEB is tiny (4 g), the TB slightly larger (6 g), and the TLNB larger still (13 g).

Telling one species of bat from another can be a challenge. The level of difficulty partly reflects the number of species you might encounter. In Hawaii there is one living species of bat. You always know what you have seen, even if you never had one in your hands. In Colombia and some neighboring countries, often there are over 100 species of bats, making identification a more difficult task. Learning to identify a bat means knowing what to look for (e.g., ears, tragus, hind foot, calcar, tail, noseleaf), what to measure (length of forearm, body mass), colour and facial patterns.

For the most part, it is 'easy' to distinguish male from female bats. Males typically have a distinct penis, which is of course not present in females.

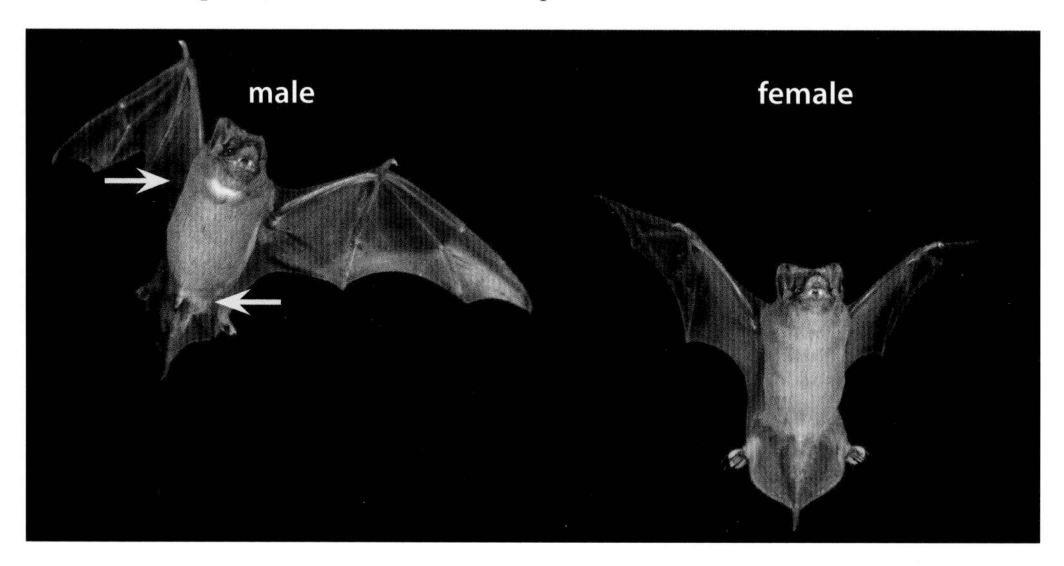

Above: Male and a female Velvety Free-tailed Bats (13 g) taking off by dropping straight down from the entrance to their roost in a building in Jamaica. The penis and the white mark on his chin, a gland, identifies the male. He uses the secretions of the gland to mark females and his spaces within the roost. Right: The worn area around her nipple identifies a nursing female Big Brown Bat.

Obtaining many details about bats means being able to recognize individuals. From the early 1930s, biologists applied numbered bird bands to bats. A band is a passive tag. To read the number you need to have the band and, with luck, the banded animal.

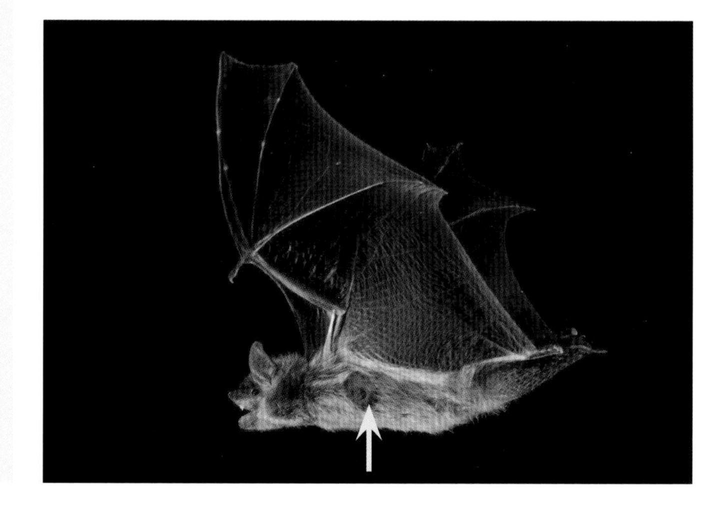

Nathusius' Pipistrelle. *JR*

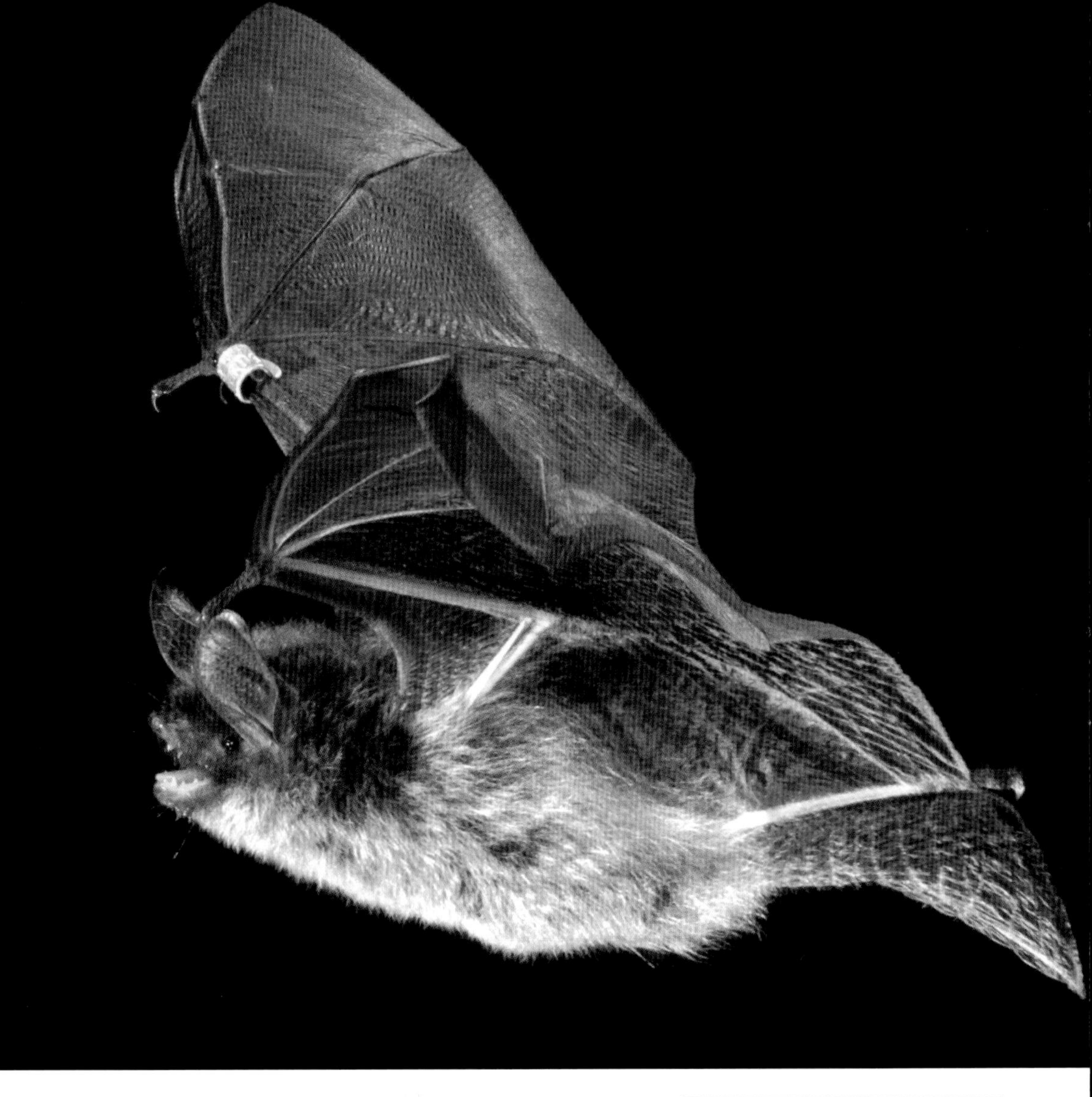

This Northern Long-eared Bat (6 g) is wearing a numbered aluminium band on its right wing. Also shown are three radio-tags we used on Black Mastiff Bats (see page 77). Opposite top: a Big Brown Bat with a Motus tag and its 19 cm long antenna. Initially we used Skin Bond® cement to attach the tags, but the bats groomed them off. We retrieved the tags from the wall in which the bats roosted and re-attached them with home-made collars made of Fibre Tape®. When we recaptured the tagged bats only one tag still had its antenna. This explained the very limited range we achieved with the transmitters. We were reminded that it's one thing to tag a bat, and another to collect data from it.

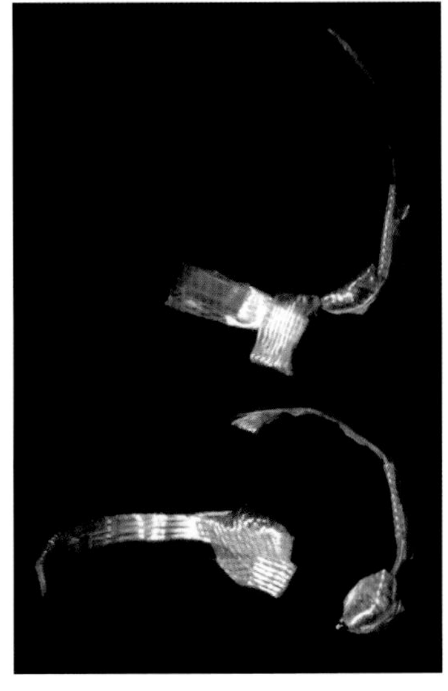

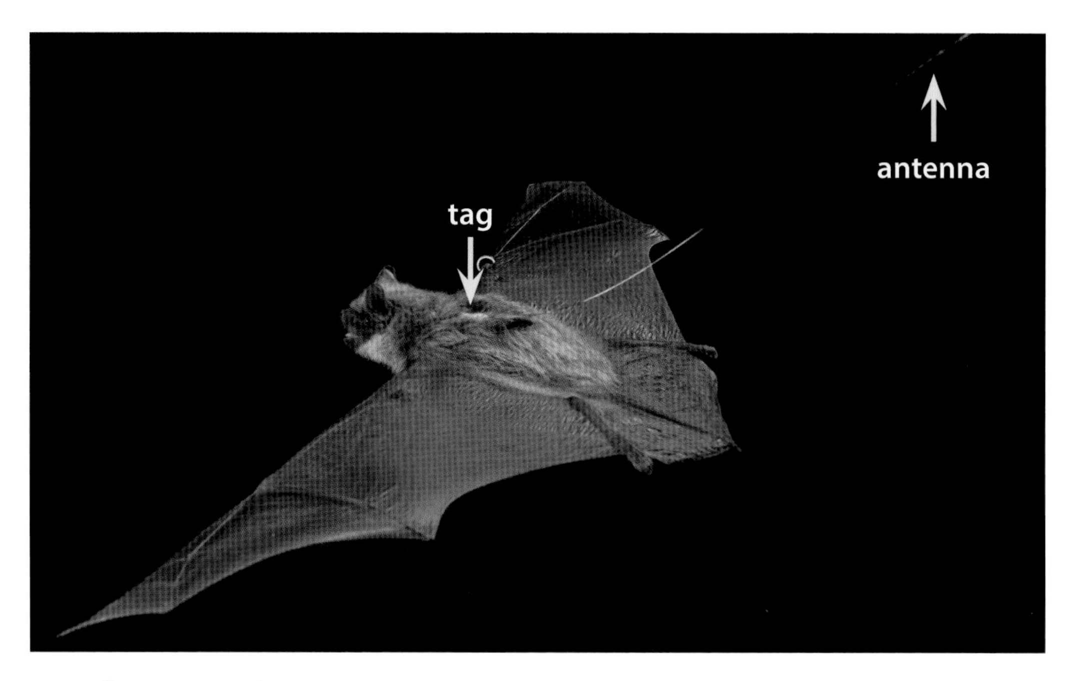

Marking and tagging

Banding studies provided information about the lifespans of tagged bats. Individuals of several species lived more than 30 years in the wild.[4] In 2005, in the Biryusa karst region of Siberia, a banded Brandt's Bat (5–6.2 g) was recovered in the field 41 years after banding. Such records are not just from temperate areas; for example, a Gervais's Funnel-eared Bat in Cuba (3 g) lived over 16 years.[5] For their size, bats are remarkably long-lived.[6]

Other band recoveries revealed movements of bats. For example, in Ontario, a male Little Brown Myotis (8 g) originally banded in December 1966 in a mine on the north shore of Lake Superior was recovered over 800 km away on 22 September 1967 at another mine near Renfrew, Ontario. On 24 October 1967, this bat was back in the mine where it had been banded.[7] Other band returns revealed that a Nathusius' Pipistrelle (~10 g) travelled 2,200 km from where it had been banded (Pape National Park in Latvia) to where it was recovered (Pitillas Lagoon nature reserve in Spain).[8]

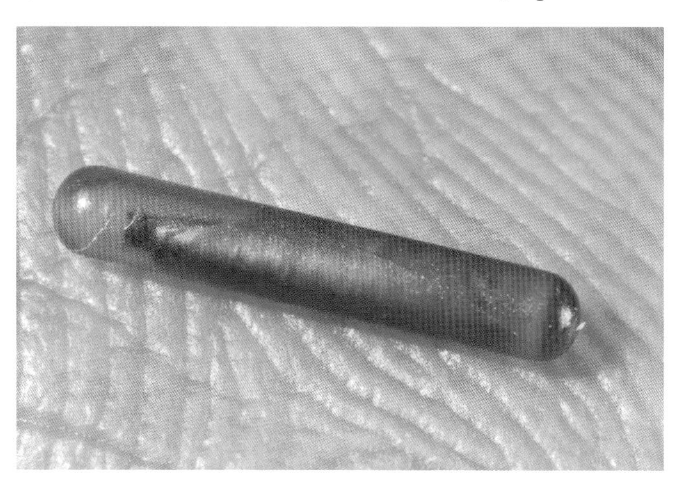

Shown for comparison is a PIT tag, 7 mm in length.

From the early 1970s bat biologists used active tags, such as radio transmitters or light tags, to follow and find bats. As technology progressed, PIT tags (passive integrated transponder) or Motus (from the Latin for 'movement') tags and even satellite tags were used to document bats' movements (see also page 163). Now identification involves receiving and reading a signal from the tag. Some modern tags will do much more, allowing the researcher to download information about where the animal has been, what it said and what it heard. Others measure proximity, when one bat is close to another (provided each is tagged). Advances in technology have significantly increased our understanding of the movements of these animals and, by inference, their knowledge of the landscape they inhabit.

As we will see (page 161), our knowledge about bats and how they move through the world has gone from a few records of banded animals to details that we did not anticipate or expect. Technology provides many wonderful tools for research. But whatever gadgets we might have at our disposal, almost invariably working with other people is key. This often means recognizing what you do not know. In fieldwork, ask the locals where they see bats, this may lead you on a merry goose chase, or to a real find. Use your imagination, know why you are looking so that you can recognize it when you see it.

Molecular biology and genetics provide an entire suite of techniques, once you obtain an appropriate sample (e.g. blood or tissue) from the bats you are studying. Some of these techniques will reveal the age and gender of the bat, what it has eaten and perhaps where it has been.

The array of possibilities is endless, and now the limiting factor, after budget, can be having the knowledge and imagination to use the tools to full advantage.

To start we will speak about how we each came to be swept up by bats.

Like other very young bats, this two-day-old Big Brown Bat has adult-sized thumbs and hind feet. It has no fur; its eyes and ears are not yet open. When they are this tiny, bands, PIT tags and Motus tags are too large, so coloured nail polish does the job for tagging.

Brock's initiation

I remember, as a child, watching a bat fly around in our cottage. When the bat landed on the stone mantle, I grabbed it, purposely going low to avoid the head. This was a mistake, but at the time I did not realize that bats hang upside-down. I was bitten and then quickly released the bat, which flew around some more before landing well out of my reach. This memory stayed with me but I learned more details only after my Aunt died. I had written to her about the adventure. Her son, my uncle, found the 'letter' among her papers and passed it on to me. This first encounter with bats occurred in July 1951.

Brock's July 1951 letter to his aunt about his first encounter with a bat.

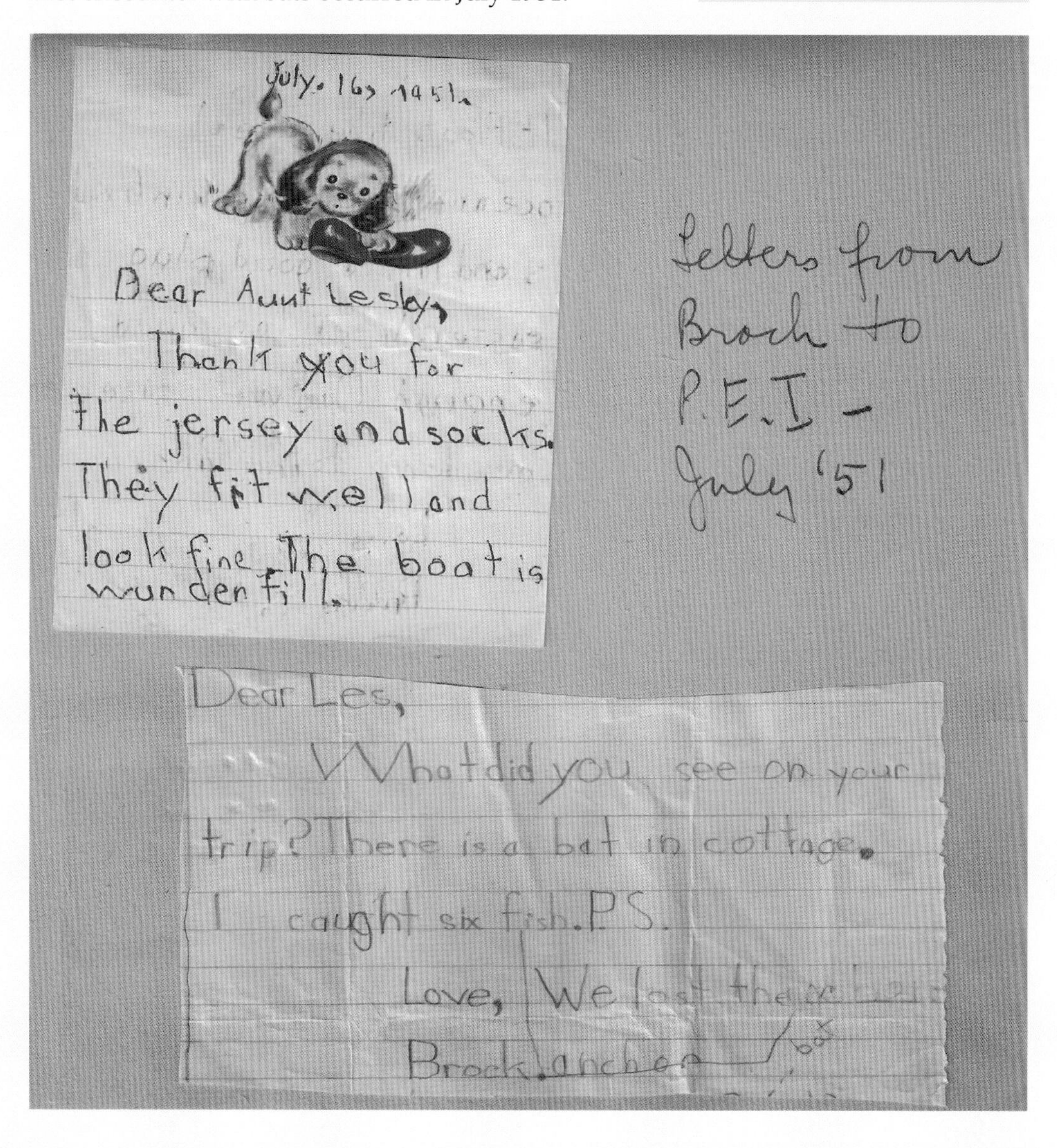

July. 16, 1951.

Dear Aunt Lesley,
Thank you for the jersey and socks. They fit well and look fine. The boat is wunderfill.

Letters from Brock to P.E.I – July '51

Dear Les,
What did you see on your trip? There is a bat in cottage. I caught six fish.
Love,
Brock

PS. We lost the [illegible] anchor

In 1963 as an undergraduate I was introduced to bats by going with a botany professor (Roland E. Beschel) to a cave to look for hibernating bats. Now the thrill of bats blended with the excitement of caves. At the same time for an Entomology course, I wrote an essay about bats and insects, drawing heavily from two books (G.M. Allen's *Bats* (1939) and K.D. Roeder's *Nerve Cells and Insect Behaviour* (1967)). What's more, I even got to hear Roeder speak and had a chat with him. Roeder had listened to bats through the ears of his moths, and he observed their defensive reactions (which also enticed Jens).

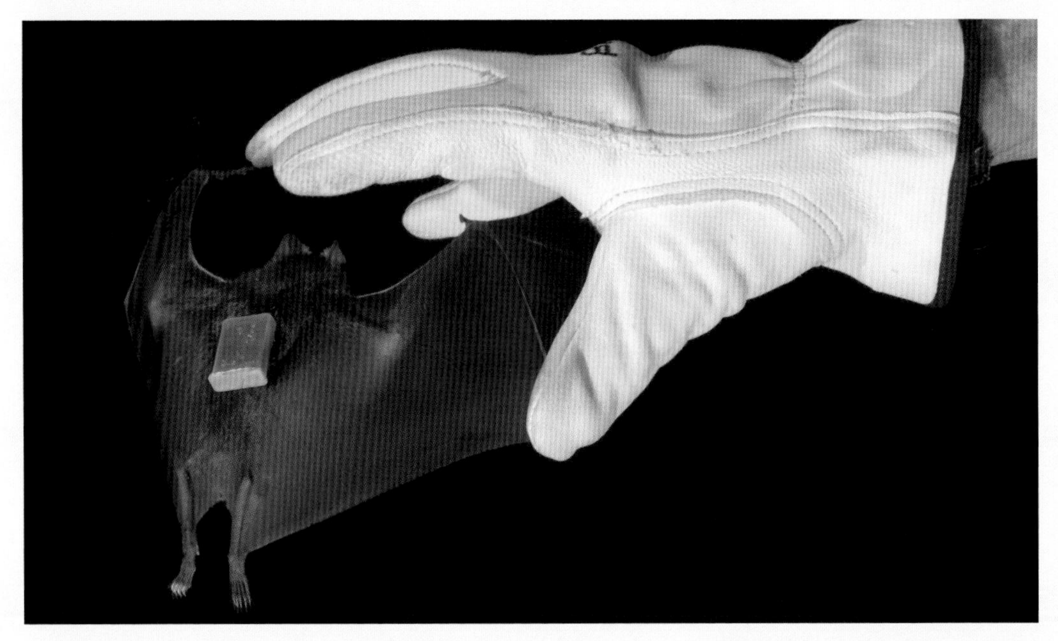

That did it and in 1965 I entered a graduate program at the University of Toronto and worked with R.L. (Pete) Peterson, then a professor at the University, and a curator of mammals at the Royal Ontario Museum. Bats had me from about 1963, and they have never let me down, although they often challenged me.

It should now be obvious that working with bats involves many different talents in addition to patience and luck. The small size of bats limits the size of tags you can use to follow them.

Bats' small size means using small tags to study them. Here MBF is gently restraining a Common Vampire Bat (30 g) carrying a GPS tag. The gloved hand tells the story while indicating the animal's size relative to the glove. To retrieve the data about where the bat had gone and when, we had to recover the tag. Our efforts were not successful because we recovered none of the 8 tags we had deployed.

Jens' start

As a teenage moth collector, I sometimes experienced competition from bats for coveted specimens circling my 125W mercury vapour lamp. A bat typically appeared at the light for a fraction of a second, grabbed a moth and disappeared into the dark. I remember how astonished I was by the elegance and speed of the bat as it swooped in under the lamp and caught the insect. Thirty years later, I learned that what I had seen at the lamp on those nights in the woods was a discovery I had not appreciated at the time.

Nothing much happened with me and bats until 10 years later, when I read a short newspaper article about a novel research project launched at Lund University. Just a telephone call and a couple of days later, I was out in the field near Lund with my about-to-be PhD supervisor Rune Gerell and a QMC bat detector. This was my first contact with the wonderful world of ultrasound and the secret life of bats. This encounter set the stage for much of my remaining life one way or another.

Moths have bat-detecting ears, allowing them to hear and evade an approaching bat. But bright lights can interfere with the moth's hearing and inhibit its evasive flights. Now the moth is easy prey for a bat, as I had noticed at my moth lamp. Normally the winner in the evolutionary arms race, the light turns the moth into a loser. In a sense, lights at night reverse the direction of evolution, and this is a very significant effect of light pollution. This tiger moth is armed with chemical defences (acquired from the lichens the caterpillars eat), and they also produce clicking sounds when attacked by a bat. The clicks, produced by noisemakers on the thorax, could warn a bat that the moth tastes bad or is poisonous. The moth clicks also could interfere with the bats' echolocation (see page 65). In the wild, Eastern Red Bats have been observed to abort their attacks on clicking moths.

A species of tiger moth which had been attracted to a light like the one Jens used to collect moths (also see page 64).

What on Earth?

Naked backs and tubular nostrils are examples of two puzzling features of bats. Each occurs in both main groups of bats (Yinpterochiroptera and Yangochiroptera). Some naked-backed bats are insectivorous, others are frugivorous (fruit-eating).

Bats' wings usually join at the side of the animal's body, but in some species they join in the middle of the back (arrow). Parnell's Moustached bat shows the typical condition, Davy's Naked-backed Bat, the alternative.

An insectivorous tube-nosed bat from Sri Lanka. This Round-eared Tube-nosed Bat (5 g) occurs from Sri Lanka to Southeast Asia. The name probably covers several related species.

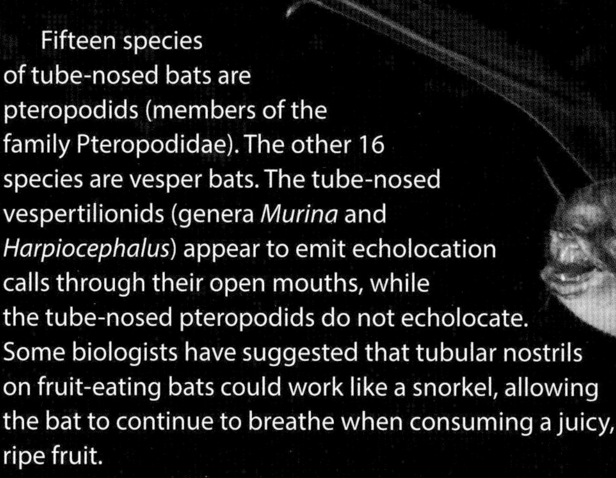

Fifteen species of tube-nosed bats are pteropodids (members of the family Pteropodidae). The other 16 species are vesper bats. The tube-nosed vespertilionids (genera *Murina* and *Harpiocephalus*) appear to emit echolocation calls through their open mouths, while the tube-nosed pteropodids do not echolocate. Some biologists have suggested that tubular nostrils on fruit-eating bats could work like a snorkel, allowing the bat to continue to breathe when consuming a juicy, ripe fruit.

In naked-backed bats the wing membranes join along the midline of the back. There is fur under the wing skin on the bats' backs.

The 13 species of fruit-eating naked-backed bats are classified in the genus *Dobsonia*. All 13 species have naked backs. These bats occur in Southeast Asia and range in size from 68 to 600 g. Unlike many other pteropodids, they often roost in caves and comparable hollows.

The three species of naked-backed insectivorous bats belong to the family Mormoopidae, genus *Pteronotus*. Big Naked-backed Bats (14 g) are larger than Davy's Naked-backed Bats (8 g). These species roost in hollows, usually in caves but also in hollow trees. Wings of the other five species in the genus *Pteronotus* join along the sides of the bats' bodies.

To date, we lack an explanation for the function of the naked backs or their evolutionary history.

2 Bat wings and flight

This pteropodid, a Wahlberg's Epauletted Fruit Bat (80 g), has its wings poised for the downstroke. Note the dog-like face, the large eyes, and tufts of white fur at the bases of the ears. Like most pteropodids, the second finger bears a claw. This species is widespread in African savannahs and often roosts in foliage (see page 131).

Bats are the only mammals capable of flapping flight. They are easily recognized by their unmistakable wings – these are in fact specialized hands made of folds of skin stretched over elongated arm, hand and finger bones. The wings attach to the sides of their bodies and their hind legs. Chiroptera, the scientific name of bats, is derived from the Greek *chiro* for hand and *ptera* for wing. The first complete bat fossils (see over) have the same wing anatomy as the 1,400+ species of bats alive today. These Eocene fossils are about 52.5 million years ago and the bones of the shoulder girdle (shoulder blades, collar bones and upper arm bones) as well as the elongated forearm hand and finger bones indicate that these bats could fly.[9] None of the living species of bats has lost the ability to fly.

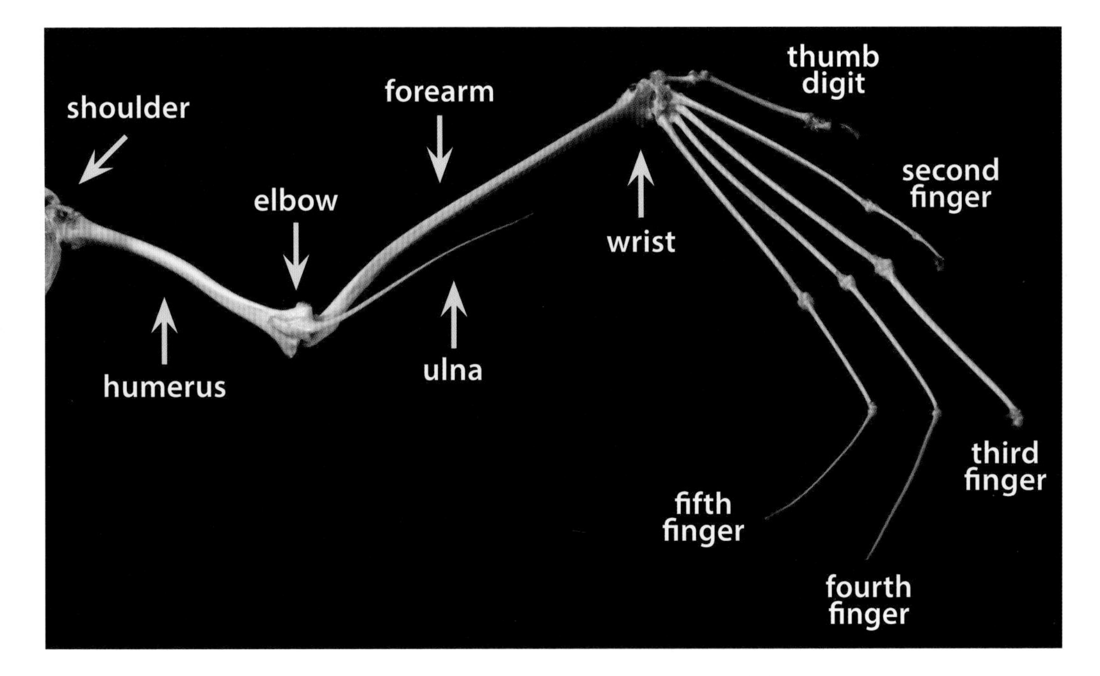

The skeleton of the right wing of a pteropodid bat. Both thumb and digit II have claws.

Fossils of early bats or of their immediate ancestors are unknown. Some fossil teeth that have been found could well have come from bats, but without the rest of the animal, important other details are lacking. The ancestors of bats were probably small forest-dwelling mammals that lived in trees and ate insects. These animals were probably nocturnal and hid by day in hollows and crevices. Various analyses suggest that bats had appeared by about 60 million years ago somewhere on the continent known as Laurasia, which comprised what is now North America and Asia.

Flying animals (bats, pterosaurs, birds and insects) flap their wings to generate propulsion, while airflow across dorsal wing surfaces generates lift. Feathers form the flight surfaces of birds, but in pterosaurs these were folds of skin, as they are in bats.

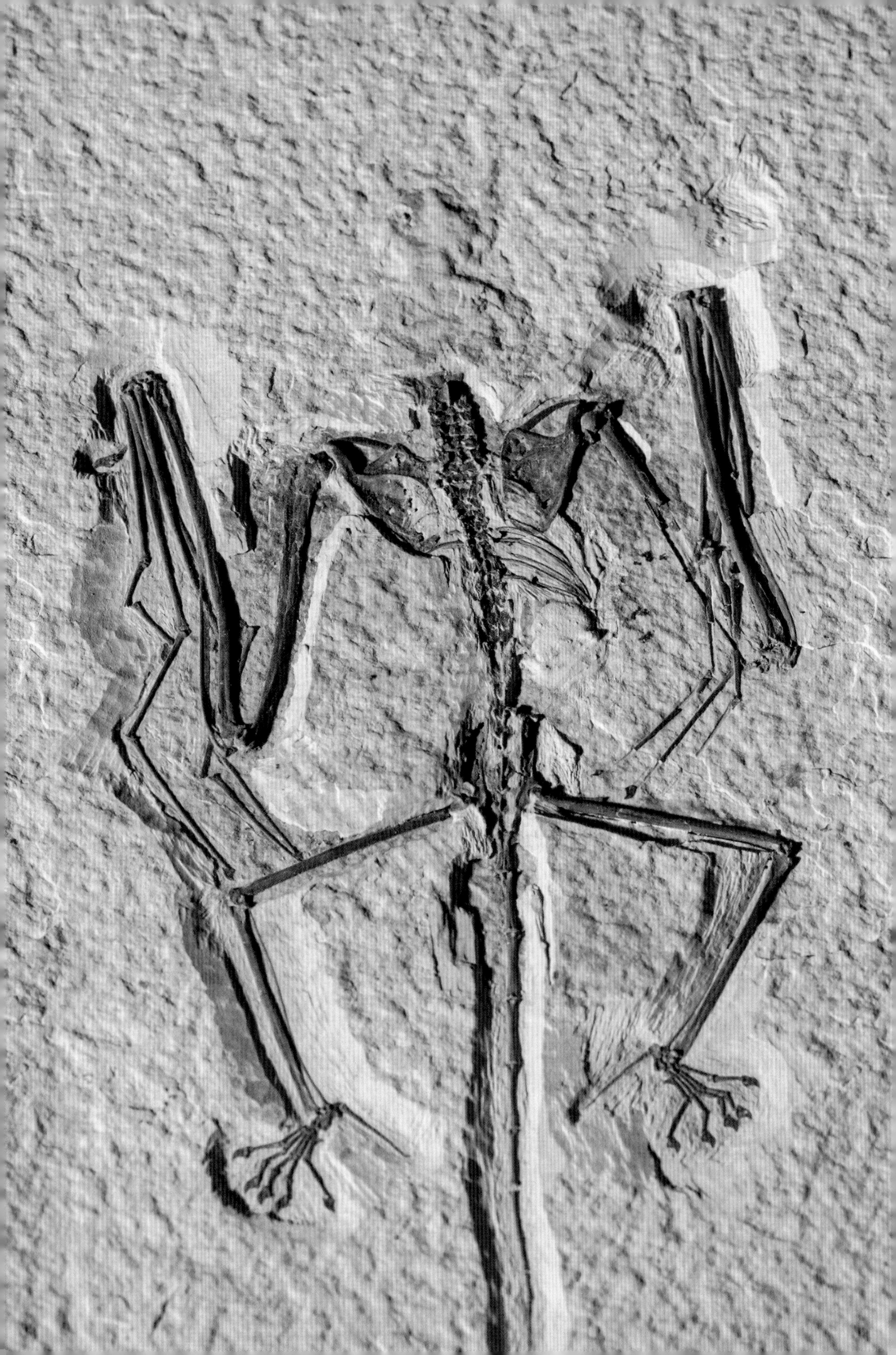

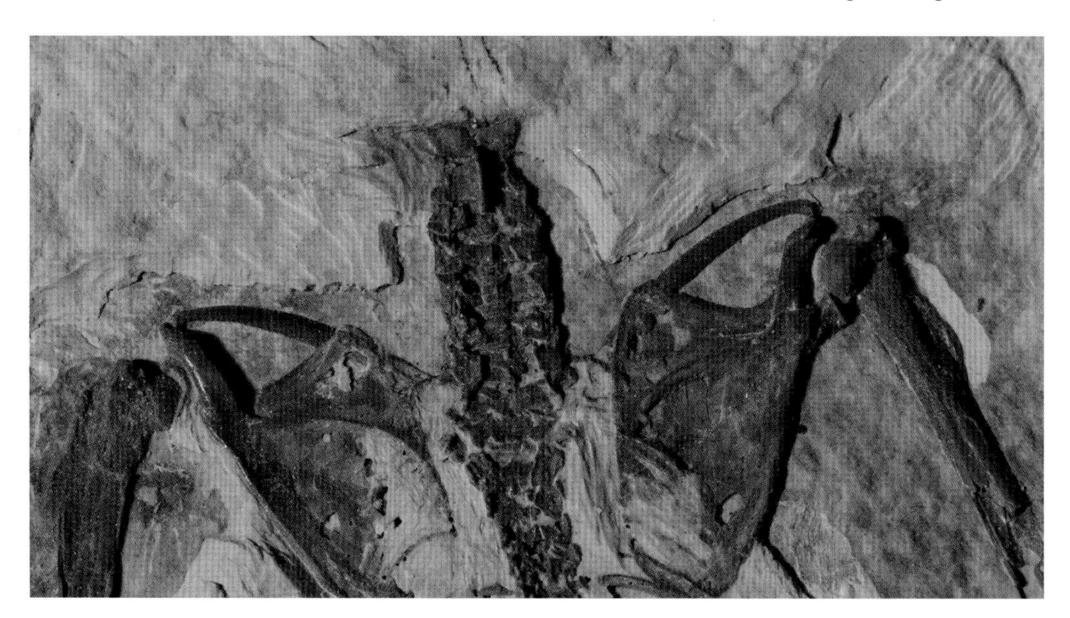

Finney's Fossil Bat, one of the oldest known bat fossils (see also page 213), from the Eocene, about 52.5 million years ago. The skeleton of the body shows the main features, tail to neck and shoulders, limbs and fingers. A closer view of the shoulder area shows the collarbones and shoulder blades, as well as the neck, some backbone and ribs. The skull and lower jaw are in a palatal view. Of special note (arrows) are the stylohyal bones (see page 50). We think this species had a forearm about 48 mm long.

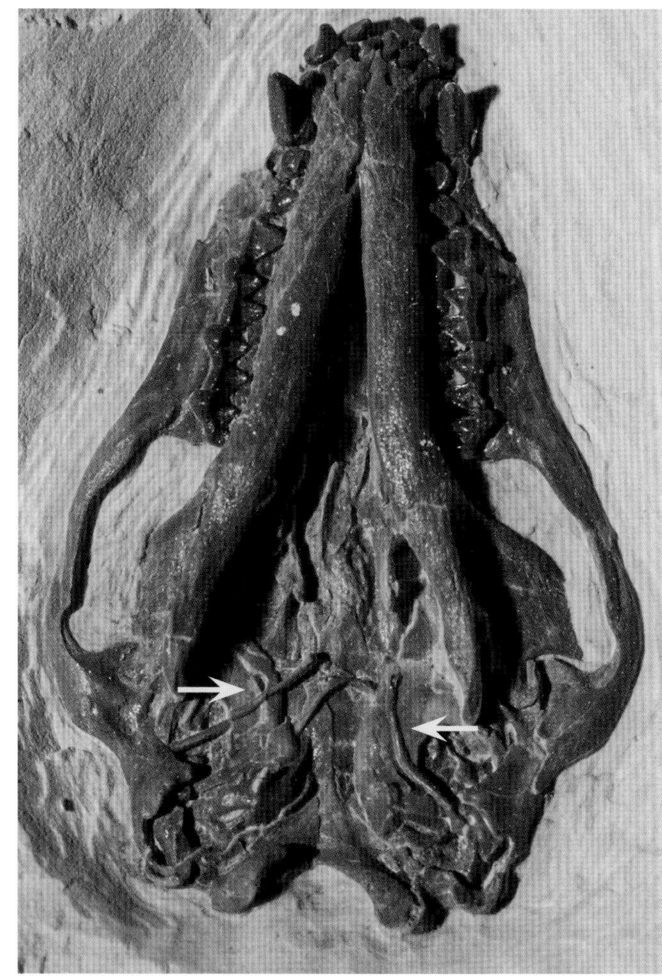

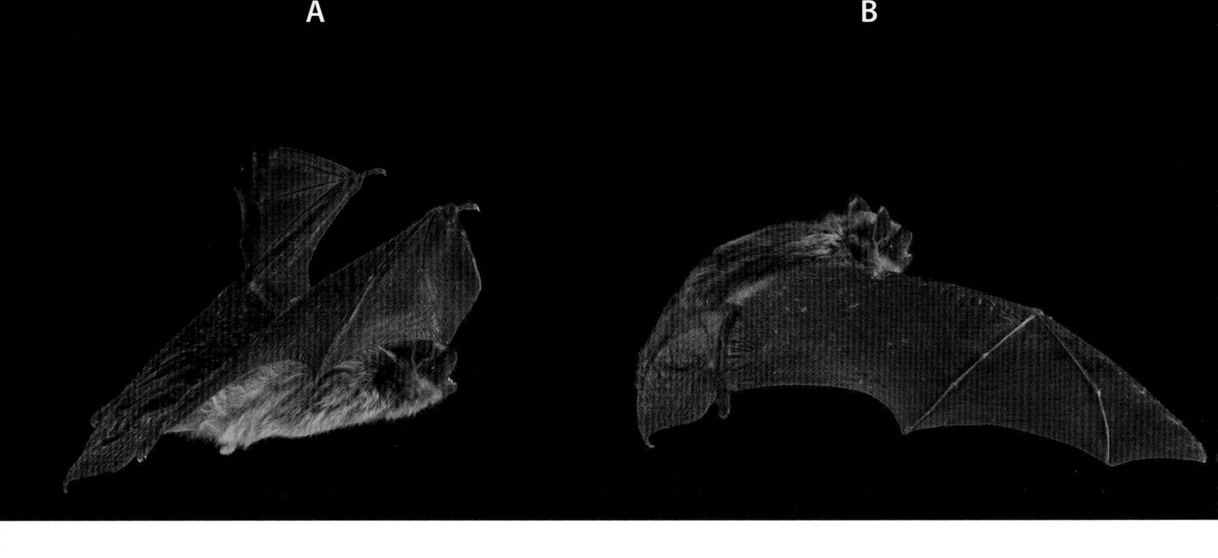

A wingbeat cycle of a Big Brown Bat flying from left to right. The bat goes from the start (A) to the bottom (B) of the downstroke, to the end of the upstroke (C) and into the next downstroke (D).

An aerofoil is crucial to flight of fixed-wing aircraft, and to the flight of birds, bats and insects. Lift is generated when air flows more rapidly across the dorsal (convex) than the ventral surface (concave) of the flight membrane. Wing movements generate propulsion. Bats and most birds typically generate most propulsive force on the downstroke. Changes in wing positions are obvious going from beginning of the downstroke through to beginning of upstroke.

Take-off and landing are two vital parts of flight. Take-off may be facilitated in bats that hang upside down. These bats need only to let go of their roost surface, drop, spread their wings and fly after attaining enough airspeed. But not all bats hang upside down. Some roost on horizontal surfaces and take off by moving to an edge and launching from there. Many species can take off from the ground. They launch by pushing off with their thumbs powered by contraction of downstroke flight muscles. Common Vampire Bats are examples of this approach to taking off. Smaller bats use the same method when taking off from water. Getting appropriate airspeed is important and explains why many bats, notably free-tailed species, need an initial freefall to build speed and fly.[11]

Bats usually make a four-point landing, wrists and hind feet touching the surface almost simultaneously. The surface may be vertical, horizontal or sloped. When landing on a ceiling, some bats approach the landing site and touch down after a somersault. When landing on leaves, Spix's Disk-winged Bats sometimes touch down feet first or make two-, three- or four-point landings. In certain cases, some bats land beside an opening and then walk into their roost. Some African free-tailed bats that roost under rocks have wart-like structures on their forearms, apparently to protect them from abrasion.

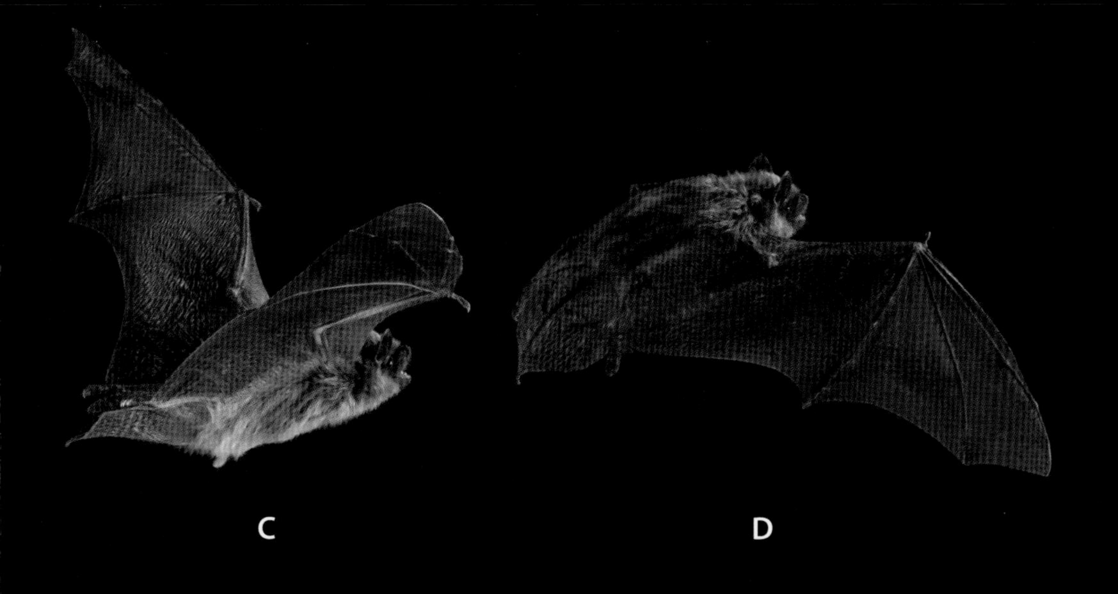

Two flashes fired 20 ms apart show a Big Brown Bat taking off from a piece of wood. The bat had just pushed off from the surface, launching itself and spreading its wings in preparation for a downstroke and powered flight.

There are tiny hairs at nodes on the ventral surfaces of bats' wings.[10] Susanne Sterbing-D'Angelo and her colleagues based at University of Maryland used an innovative approach to demonstrate that these are sensory hairs which supply bats with information about airflow over the wing. First, the researchers assessed manoeuvrability of two species of flying bats, Big Brown Bat and Seba's Short-tailed Fruit Bat. Then they used a depilatory cream to remove the hairs and again assessed manoeuvrability. Bats without hairs were less manoeuvrable flyers, but they returned to normal after the hairs grew back. The same research group also demonstrated that bat wings are very sensitive to touch, comparable to our fingertips.

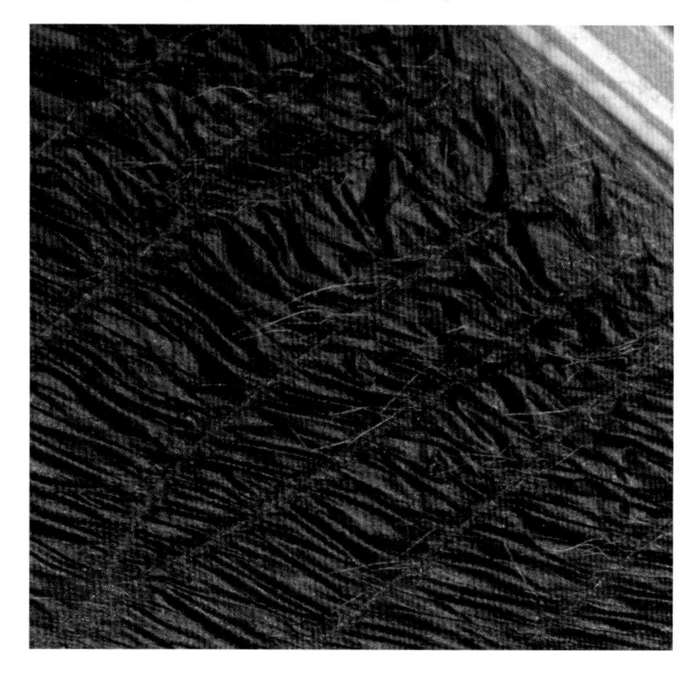

On a flying Big Brown Bat the sensory hairs show up as the grid of lines and dots radiating out from the tail towards the legs. The white dots (also present on the ventral surface of the bat's wing and tail membrane above) mark the locations of sensory hairs that provide the animal with information about airflow across the wing.

Wing anatomy

Although the skin of bats' wings is generally typical of mammals, some differences are associated with flight. These include abundant, distinctive nerve endings. The skin of bats' wings is thinner than typical mammal skin and has many bundles of fibres consisting of elastin coated with collagen. These features make the wing membrane stiffer along a front to back, compared to a side to wing tip axis. The fibres may make the wing membranes more resistant to tears. Bat wings are compliant, showing passive changes in stiffness with load, even without input from the nervous system. The compliance of bat wings may partly explain their capacity to recover from in-flight perturbations such as mid-air collisions.

Bats' wing tissues show considerable variation among species. Most species have wing membranes consisting of epidermal layers separated by connective tissue which may house nerves, blood and lymphatic vessels, sweat glands and elastin and collagen fibres. Some free-tailed bats (Molossidae) have a layer of muscle in the wing membranes. Interfemoral membranes tend to be thicker and have more layers than other bat wing surfaces.

Two bats with damaged wings. The 20 g Parnell's Moustached Bat with a healing wound on its left wing, while the 15 g Buffy Flower Bat has a striking scar on its left wing. In either case, we do not know what caused the injury.

Bats' wings are subject to damage which can leave distinctive scars. In Big Brown Bats and Egyptian Fruit Bats, wounds made by a 4 mm diameter biopsy punch healed completely in about 14 days, leaving minimal scars by 21 days. Biologists commonly use tissue punches to obtain DNA samples used in various projects. Different areas of bats' wings show different rates of healing while tail membranes heal more slowly than flight membranes.[12] Occasionally one encounters bats with actively healing wounds.

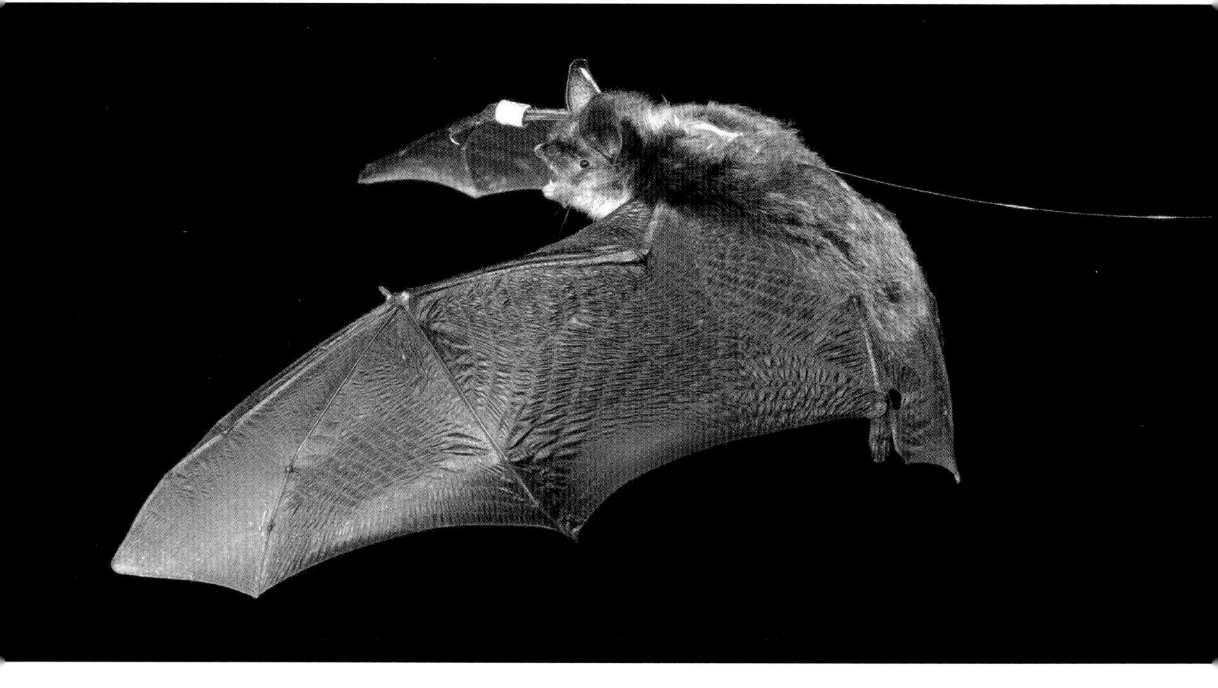

A Little Brown Myotis with a biopsy punch mark just inside its left leg in the interfemoral membrane. The bat is also sporting a Motus tag.

As resilient as their wings may be, it surely could not be possible for a bat to survive breaking a major wing bone such as humerus or forearm? It was astonishing, then, to find an adult male Big Brown Bat with a healed compound fracture of the left forearm. The bat flew readily, but with an abnormal gait, reflecting the fact that its left wing was now 11% shorter than the right one. X-ray analysis revealed a spiral or oblique fracture in an advanced state of healing. Big Brown Bats feed on the wing, so how the animal could have survived the mishap remains a mystery.

In the wings of some bats, small arteries connect directly to small veins without an intervening capillary bed. This appears to set the stage for bats using their wing surfaces as radiators, to shed body heat generated by the flight muscles.[13] In a laboratory experiment, researchers measured temperatures of Seba's Short-tailed Fruit Bats. While the animals maintained high core body (rectal) temperatures, the muscles along the upper arm did not. In resting bats, muscles along the forearms were 4–6°C (39–43°F) cooler than rectal temperatures, and 12°C (53.6°F) cooler than when the bats flew in a room at 22°C (71.6°F).

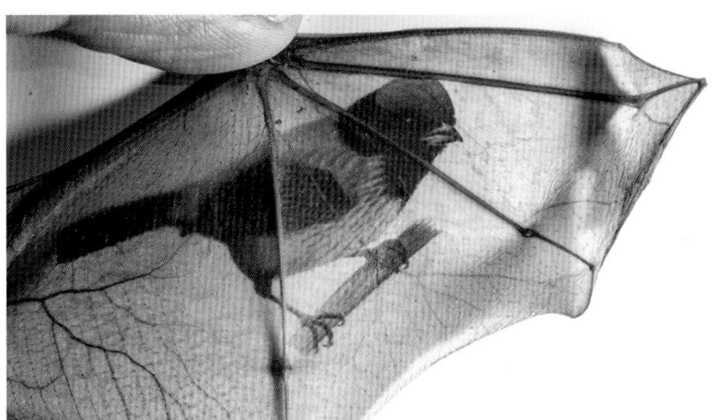

Hildegard's Tomb Bat (above, *JR*) has striking white wings. The wings of Pale-winged Dog-like Bats (left – 6 g) are translucent, even making it possible to read text or see a picture through them.

White wings

About 30 species of bats have white or translucent wings.[14] These species occur mainly in Africa, but also in South America and in tropical parts of Asia. White wings are not spread randomly among tropical species. They tend to occur in species that fly relatively fast and normally feed high in the open air. Bats that fly nearer the ground or in areas of dense vegetation have dark wings, typically shades of brown and grey. What accounts for the elimination of the dark skin pigments in white-winged bats? What harm does dark pigment do?

Dark wings could make a bat more visible at dusk and dawn, because of high contrast against the evening sky. A lighter or translucent wing might be less visible, perhaps conferring an advantage at dusk, if birds of prey hunt bats then, as many of them do. At night, in tropical darkness, do white wings confer an advantage?

The left wing of a Wrinkle-faced Bat (23 g) show more complex patterns, including a translucent area and whitish stripes.

To test the idea that white wings provide camouflage against the evening sky, we photographed black, white and transparent plastic bats against the sky and the ground. We conducted these trials in the evening, under different light intensities. The transparent bat always showed the lowest contrast against the background and the dark always showed the highest. The white bat was always intermediate. Hence, white-winged bats may be better protected from visually hunting birds than dark-winged bats. This holds at all light levels but not in darkness.

To test the idea further, we compared the activity patterns of different bat species by listening to their echolocation calls at sites in Kenya. White- or clear-winged bats emerged before any dark-winged species. These observations support the idea that the loss of the wing-pigments is an effect of risk of predation from birds at dusk. We found no obvious advantage of dark wings, at least from a predation avoidance perspective. So, why don't all bats have light-coloured wings? This is another area where we need more information about the bat behaviour. The same explanation may not apply to all species of bats that have white or translucent wings.

During a full wingbeat cycle, wing tips go from above to below the bat (see page 27).[15] Bats flying close to the ground adjust their usual wingbeat pattern, providing some benefit associated with the ground effect (see page 87).

How fast do bats fly?

Measuring bat flight speeds is challenging. The results vary considerably and depend upon how flight speeds were measured. One simple approach has been to measure the time a bat took to fly a measured distance. Researchers tend to measure flight speed in metres per second (m/s), which is not too familiar to others. Some basic equivalents are shown in the table below. Using this method, flight speeds for Eastern Red Bats were 4.3 to 6.04 m/s and for Hoary Bats 4.3 to 5.9 m/s (average 5.02 = 18 km/hr). Using a Doppler radar, we measured flight speeds of these species as they foraged in swarms of insects around lights. Now, Eastern Red Bats averaged 6.7+1.07 m/s and Hoary Bats 7.7 m/s. Silver-haired Bats carrying electronic (Motus) tags, flew 14 m/s (= 50 km/hr) when flying south crossing Lake Erie from Long Point, Ontario, to Ohio.

metre/s	kilometre/hr	feet/s	mile/hr
1	3.6	3.2	2.2
5	18.0	16.0	11.2
10	36.0	32.8	22.3

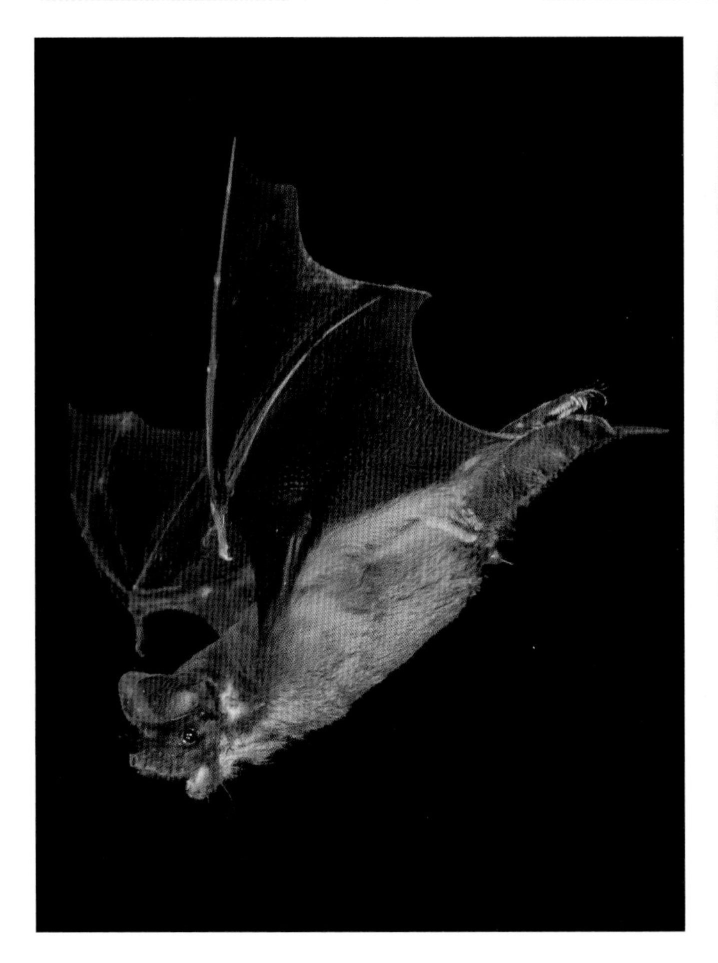

We now know that at least two species of free-tailed bats can fly much faster than previously demonstrated. First, 13 g Brazilian Free-tailed Bats (left) studied in Texas which flew from 90 to over 160 km/hr, and the 28 g European Free-tailed Bats (overleaf, *JR*) studied in southern Portugal clocked at 135 km/hr.

Until about 2015, the available data suggested that bats did not fly as fast as birds of the same size. In 2016[16] and then in 2021,[17] a couple of scientific papers demonstrated much faster flight speeds for two species of free-tailed bats. The first study involved tracking flying, radio-tagged bats from aircraft. In Texas, the tagged Brazilian Free-tailed Bats flew 25 m/s, some as fast as 44.5 m/s. The data came from seven bats flying at altitudes from 841 to 2,316 m. The 2021 study, in southern Portugal, involved 33 lactating female European Free-tailed Bats carrying GPS tags which allowed researchers to follow their movements and calibrate them relative to high-resolution wind data. These bats used updrafts to gain higher altitudes and flew up to 37.5 m/s. These flight speeds are comparable to those reported for some extremely fast birds (Common Swifts).

Drinking

Many bats obtain enough water from their food and do not need to drink. Others drink from the water's surface, on ponds, streams and lakes, and even puddles. Drinking bats use an abbreviated wingbeat cycle, stopping the downstroke and starting the upstroke when the wings are horizonal. When drinking, some bats sip , while others set up a bow wave. The ones that just plunge appear to be flying faster and are less manoeuvrable. Compare Brown Long-eared Bat (see page 55) and Savi's Pipistrelle.

A Savi's Pipistrelle (8 g) drinking at a livestock trough in Abruzzo National Park, Italy.

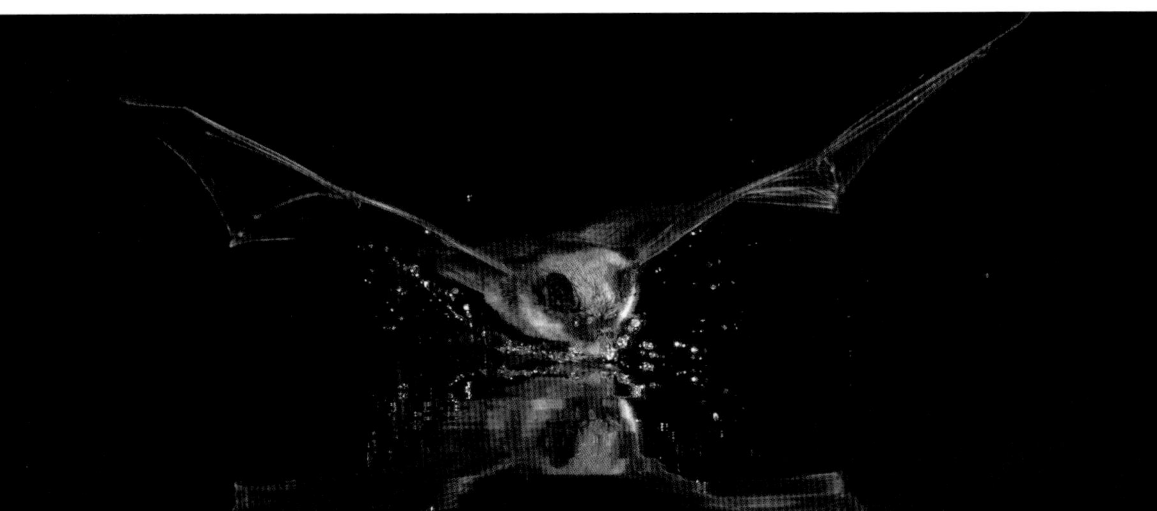

Flight antics

Our photographs of flying bats certainly produced some surprises – most notably bats flying upside down. We first-noticed this with Percival's Trident-nosed Bat (~4 g) in woodlands around a cave entrance in South Africa. These tiny insectivorous bats were the species we most often photographed, and 10–20% of the shots showed them flying upside-down. In the same wooded area, Egyptian Slit-faced Bats (10 g) occasionally (in fewer than 10% of shots) also flew upside down. In southeastern Arizona, we occasionally saw the same behaviour in a Mexican Long-tongued Bat flying upside down around a hummingbird feeder.

Two bats flying upside down, a Percival's Trident Bat (top) and a Mexican Long-tongued Bat (above).

Colour in bats

Colourful species include Decken's Horseshoe Bat (20 g). JR

Bats are much less colourful than birds. But some of the ~1,400 species are quite brightly coloured. There are other variations on this theme, such as male Eastern Red Bats which have bright red fur and black membranes with wing bones outlined in red.

Although the wings of most bats are mainly brown or black, some species have brightly coloured wing patterns, such as contrasting black and red (see page 32). In Eastern Red Bats, males are more brightly coloured than females, but other species with striking red and black patterns do not show this dimorphism (differences between the sexes). Other species, such as Painted Bats and Welwitsch's Bats, have colour patterns like those of male Eastern Red Bats. Patterns in bats tend to involve whitish or yellowish stripes on black or brown backgrounds. Pelage patterns in bats may contribute to camouflage (see page 131), breaking up the bat's outline and perhaps making it more difficult for predators to detect them.

Albinism is relatively rare in bats, and several species with white fur are not albinos. These 'ghost bats' may eat fruit (e.g. Honduran White Bats) insects (four species of sheath-tailed bats), or larger prey (Australian Ghost Bat).

In some species, fur colour varies among individuals – for example being black or red (e.g. Black Mastiff Bats, see pages 78–9). In others, fur changes from red or orange to grey or brown because ammonia, deriving from the masses of droppings in the roost, bleaches their fur.

A male Eastern Red Bat (12 g) (see page 62) is brightly coloured with striking red along finger bones highlighted by black wing membranes. Female Eastern Red Bats are much less colourful.

Other colourful examples are the Diadem Leaf-nosed Bat (45 g) (above) and the Serotine (18 g) (below), a species widespread in Europe. JR

3 Seeing with sound

A Mexican Funnel-eared Bat from Belize. These small and delicate (4 g) insectivorous bats roost in caves, often in large numbers.

A Big Brown Bat (15 g) flying towards the camera.
Its mouth is open to broadcast echolocation calls.

This Big Brown Bat in flight has its mouth open, showing an impressive array of teeth. A mammal with an open mouth display is usually threatening, but not in this case. A Big Brown Bat emits echolocation calls through its open mouth, in anticipation of echoes returning from objects in its path. Echolocation – or biosonar – is an active mode of orientation in which an animal uses echoes of sounds it has produced to collect information about what's in its path.

In 1980, Donald R. Griffin holding one end of a mist net at the entrance to a cave, Christmas Pot, near Chillagoe, Australia.

Echolocation permeates the lives of many animals, but know most about it from bats. Specializations for echolocation range from neurobiology associated with the process of echolocation, to the appearances (faces and ears) of echolocating species. In 1944, Donald R. Griffin coined the term 'echolocation' to describe the acoustic behaviour of bats. Much earlier, in 1794, Italian scientist Lazzarro Spallanzani had written about his findings on orientation in two nocturnal creatures, bats and owls. In a room lighted only by a candle, owls and bats flew freely, not colliding with obstacles. In a completely dark room his bats, Common Pipistrelles, flew readily, never bumping into obstacles such as furniture or walls. To see what happened in a dark room, Spallanzani suspended small bells from the ceiling. The bats were as proficient at orienting in darkness (they never rang the bells) as they had been when the room was lighted by a candle. Owls refused to fly in the totally dark room but if forced to, bumped into furniture and walls and rang the bells. When Spallanzani blocked one of a bat's ears, it became disoriented and was reluctant to fly. When it did fly, it regularly collided with obstacles and rang the bells. In other experiments he precluded the bats' use of cues from touch or olfaction: neither treatment resulted in disorientation of the animal. From his research Spallanzani suggested that, unlike owls, bats could 'see' with their ears. However, this theory about bats was mocked by some of his contemporaries. The behaviour became known as 'Spallanzani's bat problem', which Griffin solved about 150 years later.

The latter started working with bats as an undergraduate at Harvard University in 1934. Griffin knew about Spallanzani's problem and conducted his own studies with Little Brown Myotis. He observed the same patterns of behaviour that had intrigued Spallanzani. Then he discovered that as his bats flew, they produced pulses of sound above the range of human hearing (>20 kHz = 'ultrasonic', by definition). To do this, he had borrowed a 'sonic detector' from a physicist, G.W. Pierce, and eavesdropped on his bats as they flew. Griffin demonstrated that bats obtain information about their surroundings using differences between what they say (original sound) and what they hear (returning echoes) – hence echolocation. Later experiments revealed that several local species of bats use echolocation to detect and track insect prey. The details of the story are presented in Griffin's classic 1959 book *Listening in the Dark.*

So, by 1959 we knew much more about bats and echolocation. Griffin called echolocation a 'magic well' because each time you looked at bats more carefully, you made new discoveries. We can use the study of echolocation to illustrate the diversity of bats. There are six important points to bear in mind:

First, there are striking differences in the strengths of calls used by echolocating bats. Many species that hunt flying insects use strong (intense) echolocation calls. A Big Brown Bat, for instance, produces calls that are about 130 decibels (dB), measured 10 cm in front of the bat's mouth. This is much stronger than the wail of a smoke detector (108 dB at 10 cm). Returning echoes are much weaker, depending upon the distances they have travelled. Fruit- and nectar-feeding New World leaf-nosed bats produce much quieter calls, 80–90 dB at 10 cm. The same is true of species that take prey from surfaces (aka gleaners). Griffin called them 'whispering bats' because of their quiet signals.

The calls of whispering bats are difficult to detect. To confirm that a Seba's Short-tailed Fruit Bat was calling, Griffin built a small (<1 m^3) box. Inside there was room enough for a bat perch, a bat and for Griffin to sit holding a microphone very close to the flying

A Seba's Short-tailed Fruit Bat. This New World leaf-nosed bat (one of the Phyllostomidae) is an example of a 'whispering' bat, which uses very faint echolocation calls.

bat. These bats produced pulses of sound as they flew inside the box. In this venture, the apparatus, nicknamed 'Griffin's Orgone Box', saved the day. These and other observations revealed that it was all too easy to miss the calls of some bats and appear to 'prove' that they were not echolocating.

Second, not all bats echolocate, and echolocation is not exclusive to bats. Most flying foxes and their relatives (pteropodids – the fruit bats of Africa, Asia, Southeast Asia and Australia), do not echolocate. But some other animals do, including toothed whales, certain species of birds, various species of shrews, some species of soft-furred tree mice[18] – and even some blind people.[19]

Third, echolocation is a dynamic behaviour. As they fly, bats often change the cadence of their echolocation calls.[20] When searching for prey, a Hoary Bat typically produces five calls a second but its call rate soars to 50 calls per second after it detects and closes in on an insect. As we listen more to bats, we learn that some fly quietly, perhaps when commuting along a familiar route. The higher call rate is a 'feeding buzz' as the bat attacks a flying insect. Bats also use higher pulse rates as they approach a landing place, head to water to get a drink or in some social contexts. Echolocating bats often change the design of their calls according to the situation they face.

Further, when an echolocating bat is commuting or searching for prey its pattern of call production may be consistent and synchronized with its wingbeat cycle. Sometimes it is easy to recognize when there is more than one echolocating bat in your recording.

A Straw-Coloured Fruit Bat (250 g), an African pteropodid not known to echolocate. Note the doglike face.

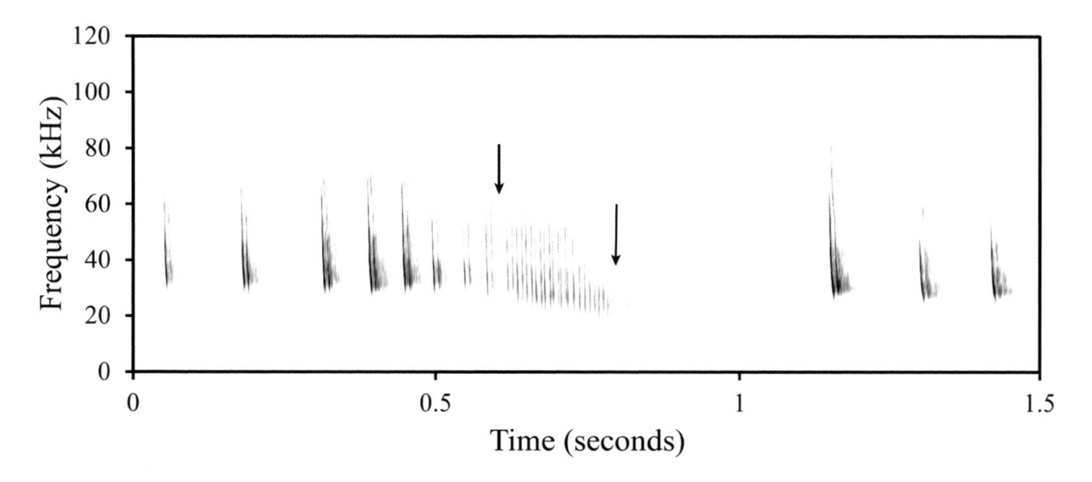

Sound pictures of a sequence of echolocation calls produced by an Argentinian Brown Bat (8 g) in 3 seconds as it searched for, detected, approached and then attacked a flying insect. Each call is tonal, starting at a high frequency and sweeping to a low one. The buzz part (arrows at start and finish) involves many calls produced in rapid succession, progressively sweeping to lower frequencies. The silent period after the buzz represents the time that the bat took to consume its prey. The vertical scale is frequency=pitch (in kilohertz, kHz), time on the horizontal axis (in milliseconds – ms) thousandths of a second.

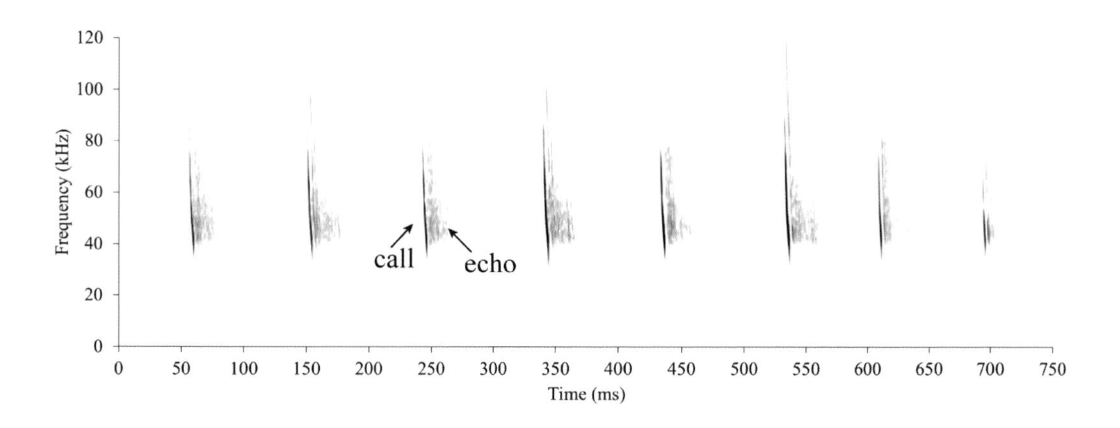

Eight echolocation calls produced by a Little Brown Myotis as it flew past a microphone. Again, horizonal scale is time (in 50 ms intervals), and the vertical scale is frequency (in kHz). Each call (dark line) swept down from about 80 kHz to 39 kHz in about 2 ms. Each call is followed immediately by an echo. Note that short calls are separated by much longer periods of silence (60 ms). This is an example of low duty cycle echolocation.

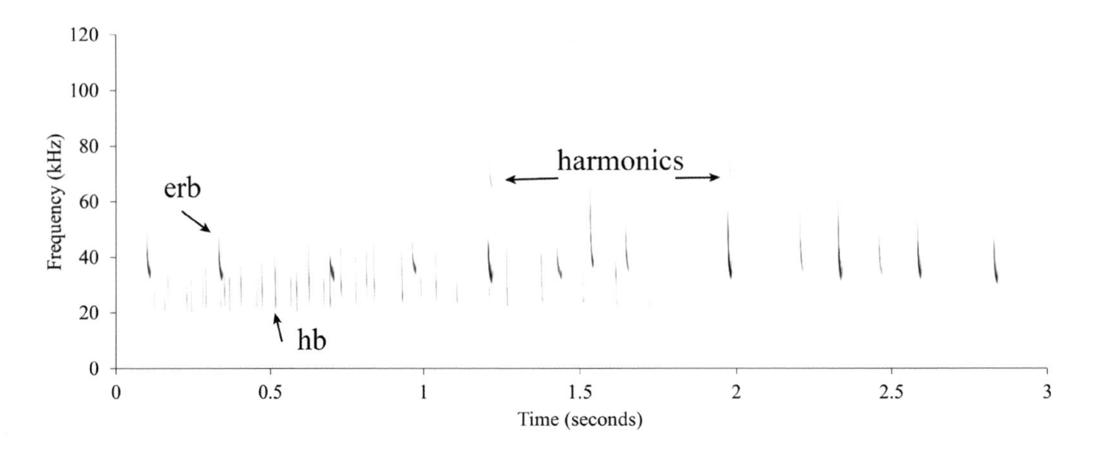

This recording shows 3.1 s with 14 echolocation calls of an Eastern Red Bat (erb), and 35 calls of a Hoary Bat (hb). The erb calls (down arrows) are stronger (blacker) and higher in frequency (50 to 37 kHz). In the background, hb (up arrows) calls are lower in frequency (30 to 20 kHz) and weaker, sometimes produced in pairs (couplets). Some erb calls have overtones (harmonics – horizontal arrows).

Fourth, echolocation is not synonymous with ultrasonic. People can hear many echolocation calls, those of birds, blind people and even some bats. Bat echolocation calls vary in frequency from low (~5 kHz) to very high (>200 kHz); ultrasonic is defined as sound frequencies above 20 kHz, the upper range of human hearing. Mind you, most people cannot hear sounds above 10 kHz.

In 1977, in Zimbabwe, we heard the echolocation signals of a free-tailed bat. We did not know which species this was but its calls, including feeding buzzes, were loud and conspicuous. In 1979 in southern British Columbia (Canada) we kept hearing the calls of another bat we could not catch. Almost a year later we realized that we had been listening to Spotted Bats (15 g), previously unknown in Canada. Local naturalists recognized the sounds but did not associate them with bats, from whom they expected ultrasonic calls.

Fifth, bats produce echolocation calls in different ways. Most species produce the signals in the voice box (larynx) = laryngeal echolocators. The few pteropodids that echolocate use tongue clicks = oral echolocators. The calls of laryngeal echolocators are tonal, showing structured changes in frequency over the call – like a whistle. Clicks of echolocating pteropodids (and those of echolocating blind people) are not tonal – they do not show changes in frequency over time but cover a range of frequencies. Hear the difference… whistle a note, and now click your tongue

Sixth, not all bats emit their calls the same way. Many species of bats fly with their mouths open (e.g. Big Brown Bats, see page 39), emitting echolocation signals through their open mouths (= oral emitters). Some other species fly with their mouths tightly closed, emitting echolocation signals through their nostrils (= nasal emitters). Some nasal emitters eat mainly insects, others mainly fruit. Most nasal emitters have leaf-like structures around their nostrils. The list includes horseshoe bats, roundleaf bats, trident bats, false vampire bats and New World fruit bats (phyllostomids). Not all

A Pygmy Fruit-eating Bat (12 g) flying with its mouth closed and emitting echolocation calls through its nostrils.

phyllostomids are nasal emitters: some, perhaps most, fly with their mouths slightly or wide open. Pteropodids that echolocate (like Egyptian Fruit Bats, see page 47) emit their clicks from the sides of their mouths.

Yet just recording (hearing) calls from an animal does not prove that it is echolocating, just as not detecting calls does not prove that it is not. Calls are but one part of an echolocation system; another element is how animals use calls and echoes to their advantage. Perhaps most telling in this context is that Spallanzani and others revealed that when deprived of the ability to hear echoes of their calls, some bats were grounded.

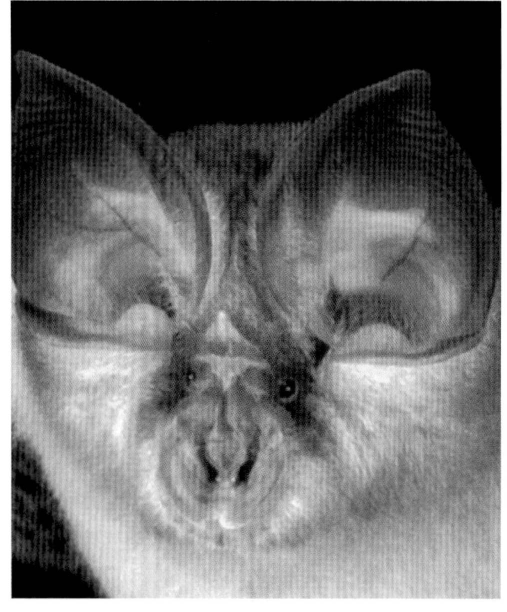

Rufous Horseshoe Bat

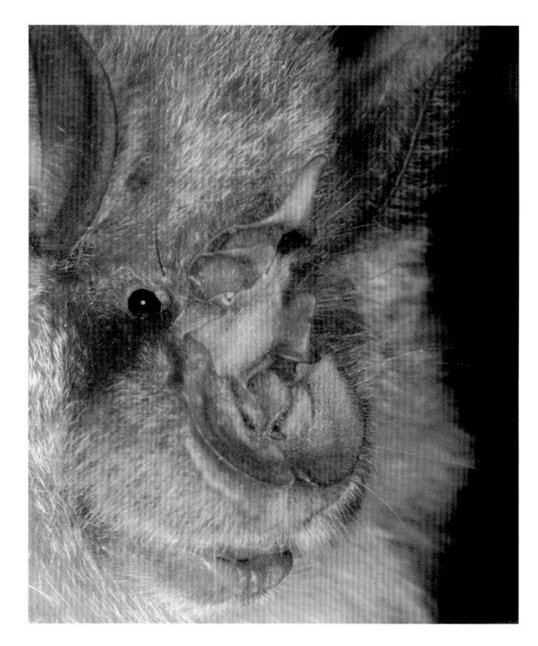

Bushveld Horseshoe Bat

Schneider's Leaf-nosed Bat

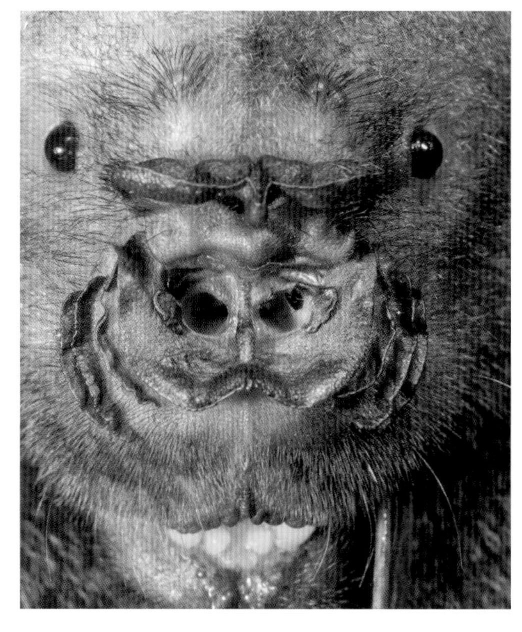

Indian Roundleaf Bat

Four insectivorous bats from the Old World with noseleafs. All emit their echolocation calls through their nostrils.

Bats with noseleafs are said to fly with their mouths closed. A Pale Spear-nosed Bat demonstrates this. But a Pallas's Long-tongued Bat flying directly at the camera has its mouth open with its tongue protruding.

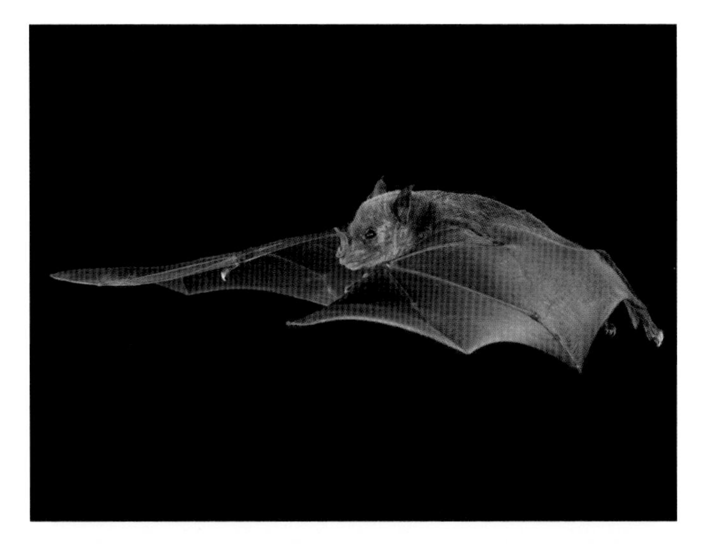

An Egyptian Fruit Bat (150 g) emits its echolocation clicks from the side of its mouth (arrow).

The perils of generalization

Phyllostomid bats, of which there are about 150 species, show exceptional diversity even for bats. Diet is an indicator: while some eat fruit, others feed on nectar and pollen, still others consume animals, from insects to other bats. The most notorious are the blood-feeding vampires. The facial appearances, behaviours and diets of these bats remind us about the folly of generalizations. New World leaf-nosed bats (family Phyllostomidae) were said to be nasal emitters of echolocation calls, and indeed some are. Photographs also revealed that some species of phyllostomids consistently fly with their mouths tightly shut. Still others fly with their mouths wide open. A further variation is that some species fly with mouth partly open and tongue protruding. An image of a bat with its mouth open does not prove that it is emitting echolocation calls orally. Do some species switch back and forth? This leads to another question: what role might the tongue play in echolocation?

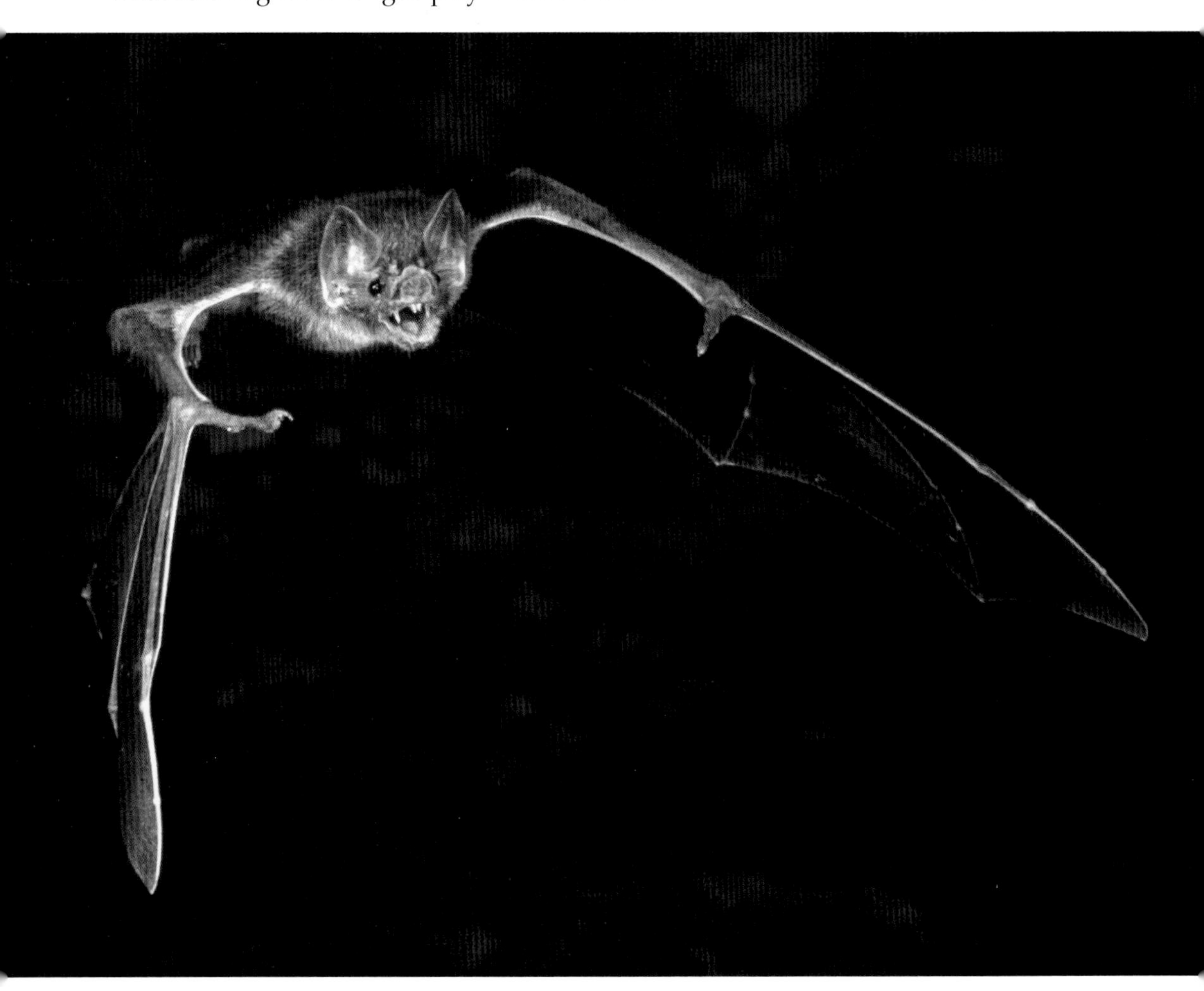

A Common Vampire Bat (30 g) emerging from a tunnel roost. Note the wide open mouth, the long, prominent thumbs and the distinctive teeth (upper incisors and canines).

Basic echolocation

Bats echolocate by using the differences between pulses and echoes. To make the comparisons, the bat must register outgoing signals in its brain for future comparisons with returning echo(es). Most echolocators (and most bats) reduce the chances of self-deafening by separating pulse and echo in time. Strong outgoing signals mask weak returning echoes, so these bats cannot broadcast and receive at the same moment. This explains why bats produce shorter and shorter calls as they close in on insect prey. Specifically, echoes return faster and faster as the bat approaches its target. In engineering terms, these echolocating bats are low duty cycle, because they produce short calls separated by long periods of silence.

Other bats (about 200 species from four families) separate pulse and echo by frequency. They can simultaneously broadcast and receive, exploiting Doppler shift – the changes in frequency associated with movement of the target (an insect) relative to the source (the bat). Lowering the frequency dominating each outgoing pulse alters the frequency dominating the echo. This 'Doppler shift compensation' makes these bats very good at detecting the flutter of a flying insect.

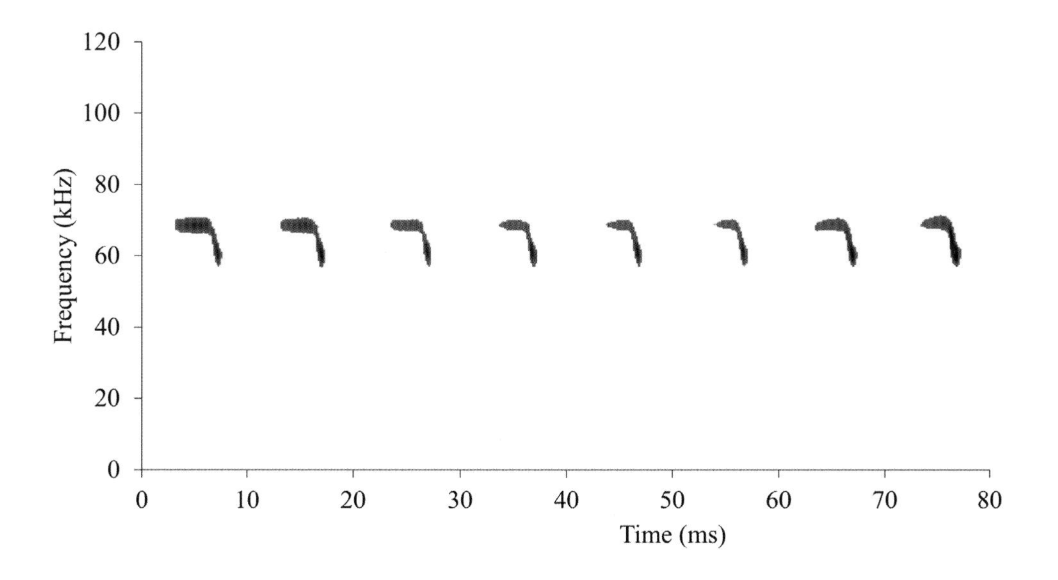

These eight calls were recorded from a Mesoamerican Moustached Bat, a species that separates pulse from echo by frequency. Here long calls are separated by shorter (compared to low duty cycle) periods of silence.

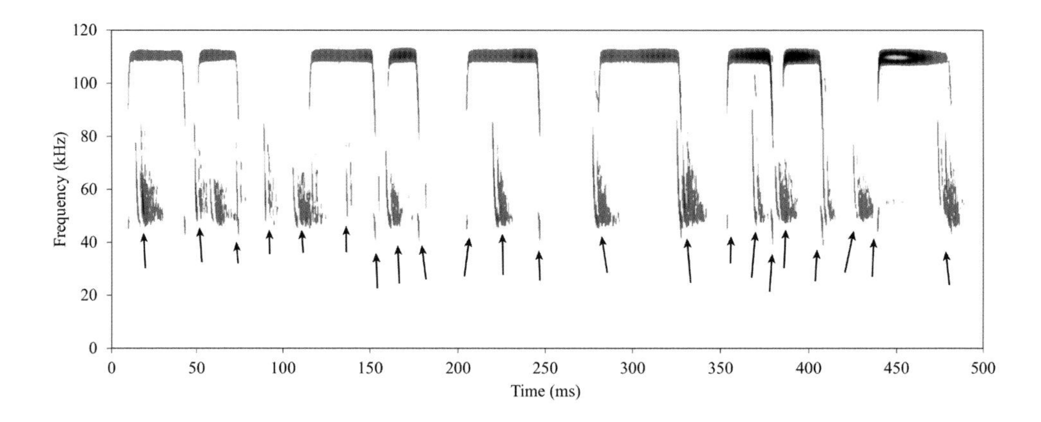

A Lesser Horseshoe Bat and a Bodenheimer's Pipistrelle were observed flying in the same airspace at the same time. In the resultant spectrogram, the nine horseshoe bat calls are across the top, and 15 pipistrelle calls (arrows) below.

In engineering terms, these are high duty cycle bats, using long signals separated by short periods of silence. Horseshoe bats, roundleaf bats, trident bats and a few species of moustached bats employ this approach – all of them use long echolocation calls dominated by a single frequency. Some, horseshoe bats, roundleaf bats and trident bats have noseleafs but moustached bats do not. Noseleafs influence the outgoing biosonar signals, including their directionality and the presence of harmonics or overtones. But noseleafs also occur in low duty cycle bats.[21]

Why echolocate?

Echolocation is expensive: there are significant energy costs to producing signals, receiving echoes and then processing the data. But echolocation allows animals using it to operate under conditions of low or uncertain lighting, or in the dark. One reflection of advantages is the variety of animals that echolocate.

When did echolocation first appear? Did Finney's Fossil Bat (see pages 24–5) echolocate? We do not know because to date we do not have hard evidence (bones) always associated with echolocation behaviour. What about the immediate ancestors of bats? Bat biologists who think that echolocation was a key to bats' origin and success would say that early bats did use it. Echolocation confers a big advantage, namely access to flying nocturnal insects as food. This important food resource was not available to other flying vertebrates such as birds, which appeared several million years earlier than the first bats. Echolocation allows modern birds that have this facility (cave swiftlets and the oilbird) to nest in the darkness of caves, perhaps reducing some predation on eggs and nestlings. So far, there is no evidence that echolocating swifts use echolocation to find prey.

Modern bats producing echolocation calls in the larynx have a direct bony connection between the auditory bone and the voice box.[22] Specifically, the stylohyal bone contacts (and may be fused to) the tympanic bone. This connection permits the registering of

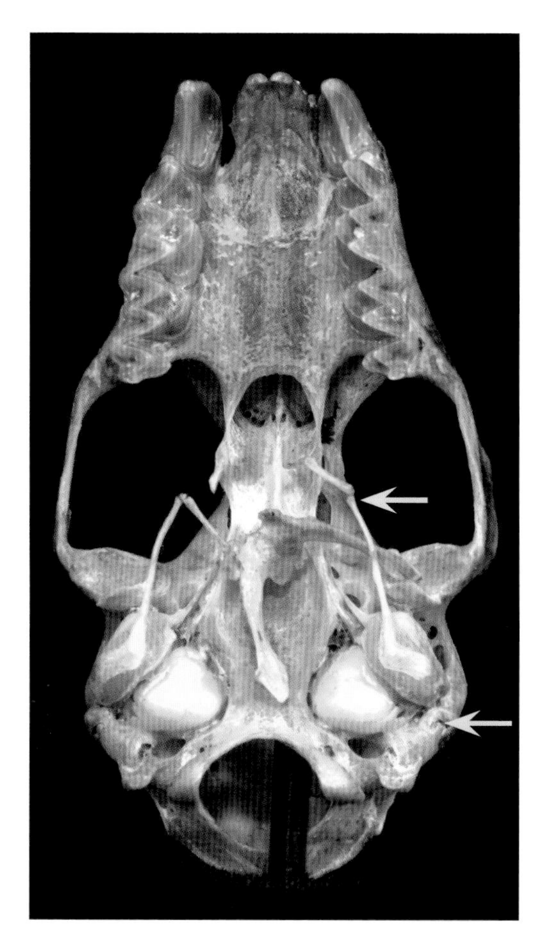

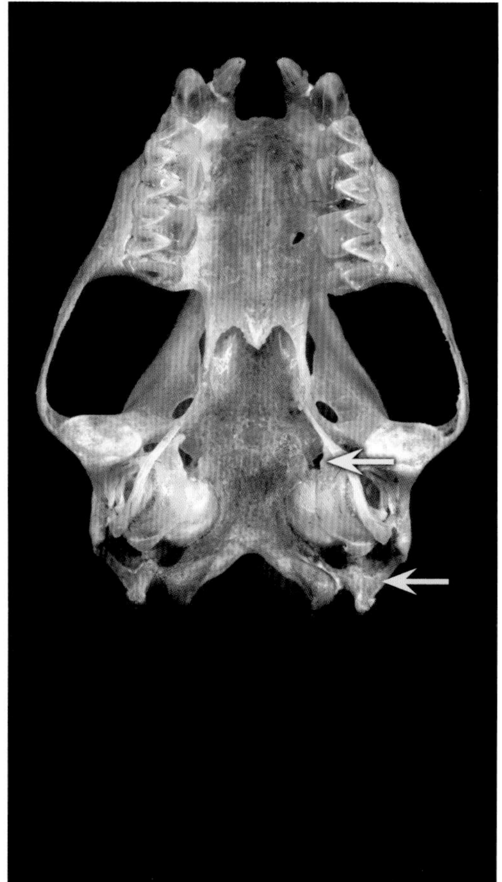

A palatal view of the skulls of a Great Roundleaf Bat (left – 60 g) and a Harlequin Bat (right – 6 g). In each bat, arrows show the stylohyal bone connecting the tympanic bone (bottom) to the hyoid arch (top). In the Harlequin Bat, the stylohyal bones are less firmly attached to the hyoid bones than in the Great Roundleaf Bat.

an accurate representation of each outgoing signal (what the bat uttered) in the bat's brain. This sets the stage for comparing pulses and echoes. In bats, as in other modern mammals, the three bones of the middle ear (malleus, incus, stapes) are derived from bones in the lower jaws of their ancestors. In some mammals, chewing interferes with listening. Some bats can echolocate and chew at the same time, which is facilitated by the stylohyal connection to the tympanic bone. So, what do the fossils say? Finney's Fossil Bat has conspicuous stylohyal bones (see page 25), but the fossils are 'pancake' – too flattened to reveal if they contacted with the tympanic bones. In pteropodids, including those that echolocate with tongue clicks, stylohyals do not contact auditory bones. In 2022, R.B. Sulser and colleagues used previously unnoticed neuroanatomical evidence to support the argument that laryngeal echolocation was the ancestral condition in bats.[23] These findings added weight to the idea that pteropodid bats have lost this ability.

Echolocation and the faces of bats

Bats' facial features and ears are integral parts of the echolocation system. Patterns of call emissions and beam width involve some combinations of mouth, noseleaf, lips and grooves. Ears, technically known as pinnae, are key to reception of echoes. The sizes and shapes of the facial structures and ears reflect the wavelengths of emitted signals and returning echoes. Echolocation is dynamic: bats change the shapes and sizes of their mouths, noseleafs and ears to maximize the information available from echolocation, partly by influencing the sound beams they emit.

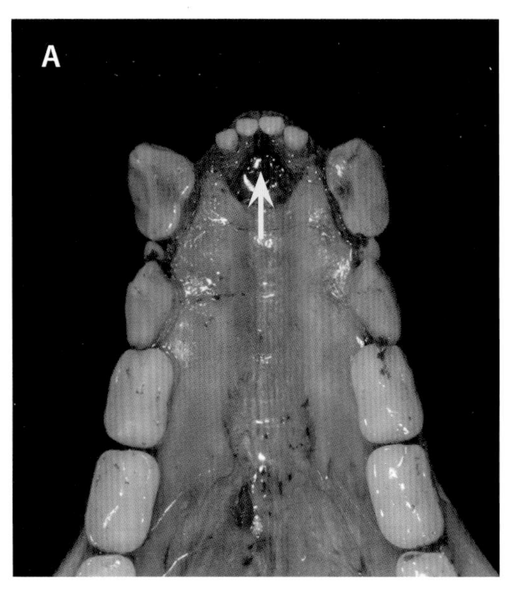

Geoffroy's Rousette

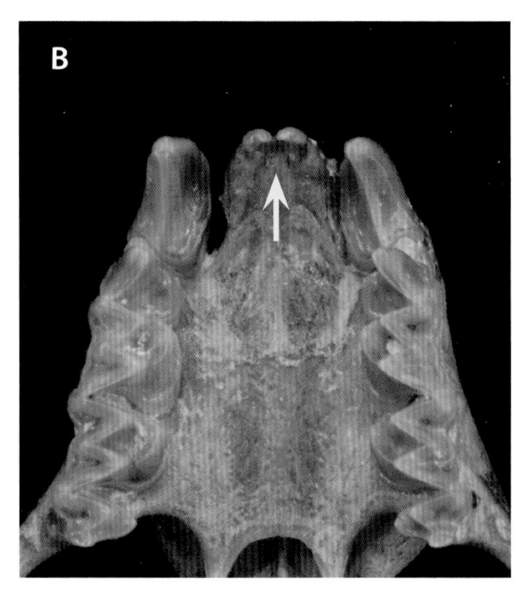

Great Roundleaf Bat

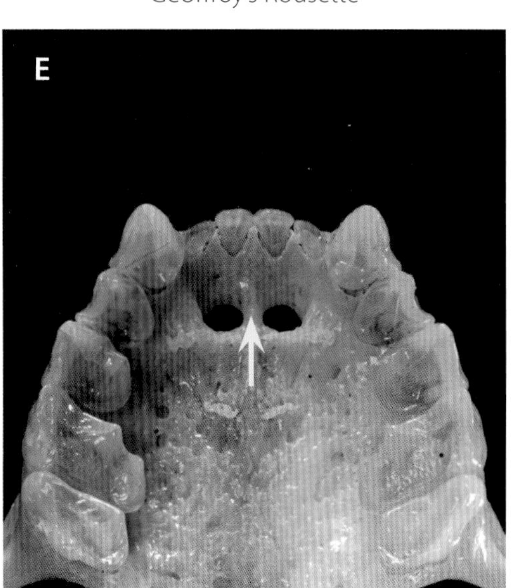

Great Fruit-eating Bat

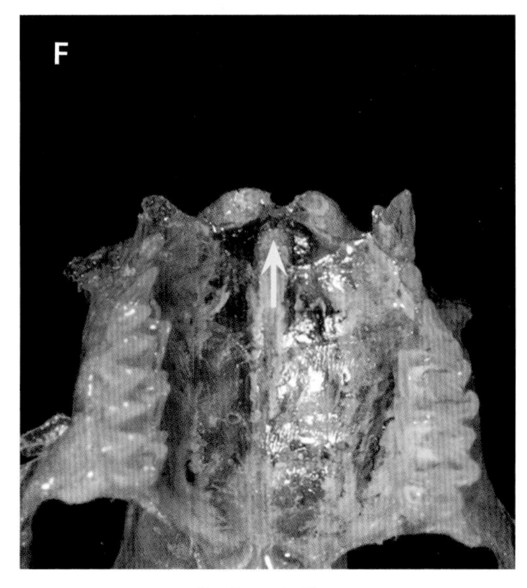

Proboscis Bat

Variation in the anterior palates of bats (arrows). In A, C, D and E, obvious premaxillary bones extend from canine to canine and bear upper incisors. In B, premaxillaries extend out from the palate but do not extend from canine to canine. In F, there is a gap behind the premaxillaries, in G and H an open space between upper incisor teeth.

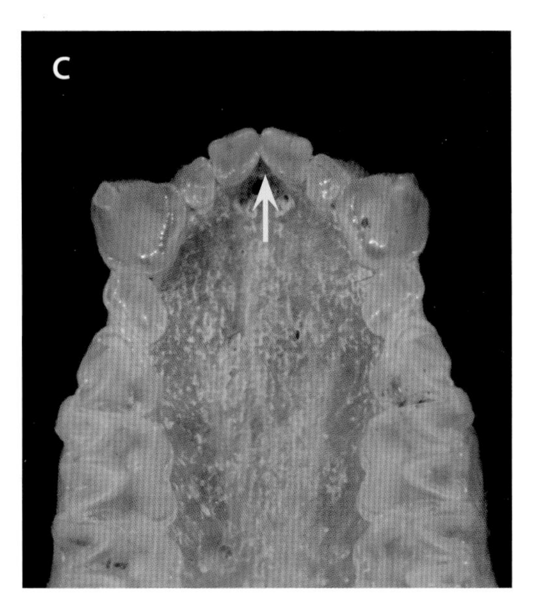

Parnell's Moustached Bat

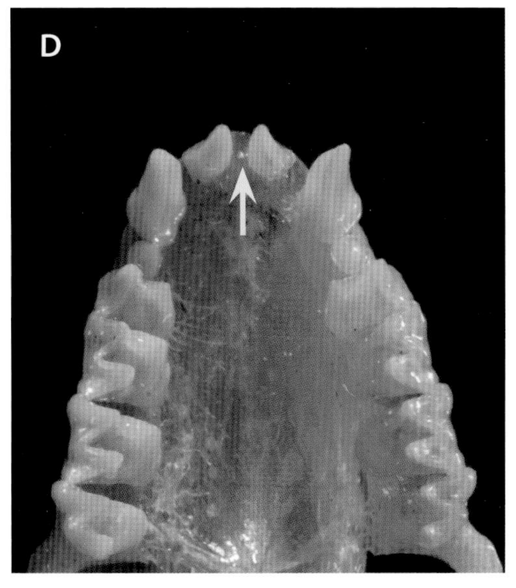

Large-eared Free-tailed Bat

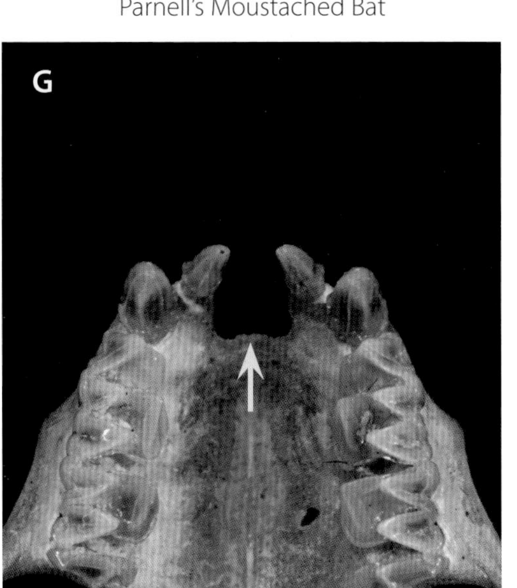

Big Brown Bat

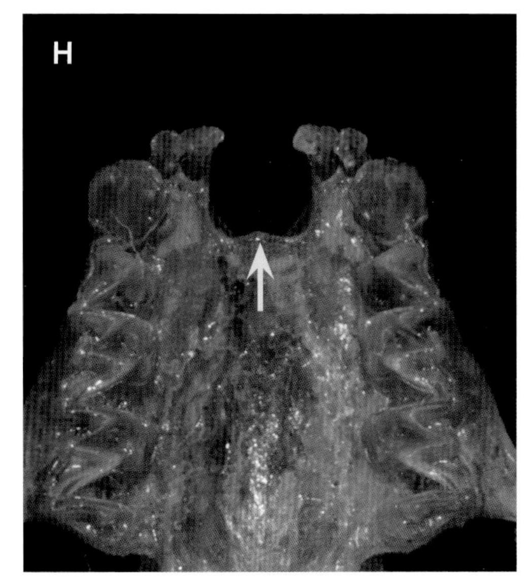

Harlequin Bat

Bats show considerable variation in palatal structures.[24] Premaxillary bones typically bear upper incisor teeth, affecting bite strength and influencing flexibility of the anterior part of the palate. Premaxillary bones are reduced or absent in at least eight lineages of bats, so the anterior part of the palate is flexible. A reduction in the premaxillaries is obvious from the gap between the upper incisor teeth. Flexibility of the anterior palate allows Big Brown Bats to change the shape of their open mouth, adjusting gape to control beam shape and width according to the situation. Desert Pipistrelles and Daubenton's Bats are other species that have flexible anterior palates. There are further ways to achieve beam control (pages 58–9).

Two views of a flying Big Brown Bat reveal changes in the bat's upper lip. In the left view the upper lip is raised, revealing most of the bat's upper left canine tooth. In the right view, only a small portion of this tooth is visible. The face-on skull below, shows the gap between the upper incisor teeth.

Bats' external ears can be large and conspicuous.[25] Ears collect soundwaves from the air, and their size and shape reflect both wavelengths and direction from which sounds emanate. Bats that eat other creatures can use their pinnae to collect echoes from their calls and/or sounds produced by potential prey. Large pinnae can make bats more sensitive to faint sounds – so, when hunting, long-eared bats usually hold their ears up, facing the flight direction. While commuting,

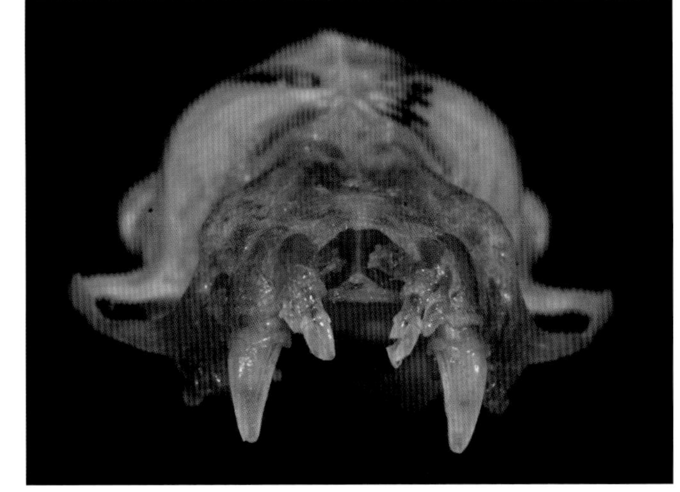

the same bats tilt their heads forwards so that the ears are almost horizontal.[26] This change in position and attitude can reduce drag and generate lift, making flight less expensive. Now, instead of hindering flight, the ears serve as an extra pair of wings, almost like the forewings of a modern fighter aircraft.

Brown Long-eared Bat flying with ears up as it sips water.

Brown Long-eared Bat flying with ears cocked forward. *JR*

Among free-tailed bats, many species have large ears, coinciding with echolocation calls dominated by lower frequency sounds (< 20 kHz) which travel farther in air than higher frequencies. These bats fly fast and often feed at higher altitudes, where they must scan large volumes of air to detect echoes from insects. Free-tailed bat species that have big ears cannot change their ear positions as much

The ears of a Dwarf Bonneted Bat (12 g) join over his muzzle.

as other long-eared bats. The ears of some Large-eared Free-tailed Bats often join atop the muzzle, perhaps generating a certain amount of lift during fast flight.

Many bats have pinnae with uniformly spaced horizontal ridges, folds or arrays of grooves.[27] In some species, folds coincide with a bat's ability to fold back its ears. This may permit control of the size and spacing of the grooves. A comparison of 120 bat species revealed a range from zero up to as many as 20 grooves, but provided no clear pattern with respect to sound reception or behaviour. Variation in grooves appeared to reflect hunting behaviour, with more grooves in species that searched for prey on surfaces. The function of the folds is a topic of ongoing work.

A Davis's Round-eared Bat can hold its ears erect or folded back. Note the grooves or pleats in the ear, which are more conspicuous when the ears are folded.

Beam control and bite power

The striking differences between faces and lips may well coincide with differences in echolocation behaviour.

A Parnell's Moustached Bat (20 g)

A flexible palate means better control over the sonar beam. Phyllostomid bats, and their close relatives the moustached bats, have robust premaxillae and upper incisor teeth (see page 53) as well as powerful bites. Phyllostomids use their noseleafs to control the beam of echolocation signals. Moustached bats appear to use their large lips to control beam patterns. The premaxilla and upper incisor teeth could give these bats a stronger bite than similar sized species lacking premaxillary bones. These insectivorous bats occur in the Subtropics and Tropics of South and Central America; some also occur in the Caribbean.

An Antillean Ghost-faced Bat (8 g) flying in a cave in Jamaica.

4 Echolocation: a window onto bat behaviour

Eastern Red Bats are common in the United States and Canada, roosting alone among the foliage of trees and vines. They emit echolocation calls through their open mouth.

An echolocating Hoary Bat (30 g) flying with its mouth open.

Today, bat biologists and others eavesdrop on echolocation calls to assess levels of traffic and determine which species are active in a study area. Sometimes we can identify a bat species by its echolocation calls too. Hoary Bats are a good example: they produce echolocation calls with most energy between 18 and 25 kHz. In many places throughout their range the echolocation calls of Hoary Bats are distinct, making them easy to identify. With energy below 20 kHz, Hoary Bats' calls are readily audible to children, but not to older people, who will have lost much of their high-frequency hearing. The frequencies in the bats' calls mean that biologists typically use bat detectors to monitor them. It was surprising to researchers in California to catch Hoary Bats in mist nets or harp traps without having heard the bats' calls just before capture.[28] Did silent flight mean that the Hoary Bats were not echolocating? Or were they behaving like a species of whispering bat?

Using an array of microphones at sites in Humboldt Redwoods State Park (California), researchers confirmed echolocation behaviour and recreated the flight paths of Hoary Bats.[29] Furthermore, in addition to 'normal' echolocation calls, bats flying in open habitats sometimes produced 'micro' calls. These were higher in frequency, shorter and much less intense than 'normal' calls. The weak calls would have been of little use in echolocation, allowing a bat to detect a tree-sized object only at 7.5 m, as opposed to almost 30 m with normal calls. Flying at 7 $m.s^{-1}$, this meant having 0.9 s versus 6.9 s to react and avoid a collision.

The researchers also noted a higher incidence of quiet flights during the mating season. This suggested that the bats they studied reduced their conspicuousness to other Hoary Bats at a time when males might be competing for females. Playback presentations of Hoary Bat calls confirmed that these bats were attracted to the calls of conspecifics (members of the same species).

This episode illustrates how bats might exploit the echolocation calls of others. Changing echolocation calls, as Hoary Bats did in the field, may allow them to go unnoticed by other Hoary Bats. In a laboratory, captive Big Brown Bats trained to find and take tethered mealworms display similar behaviours. When another bat was added to the scene, some individuals changed their calls to make them distinct from those of the interloper. They did this by changing the frequencies dominating the calls, the cadence and call intensities. Other individuals just flew silently. Echolocation can give the user access to resources such as food or nest sites, but such calls also reveal its presence and identity, which can represent a significant disadvantage. So, it is no surprise that Hoary Bats, Big Brown Bats and others take steps to thwart eavesdroppers.

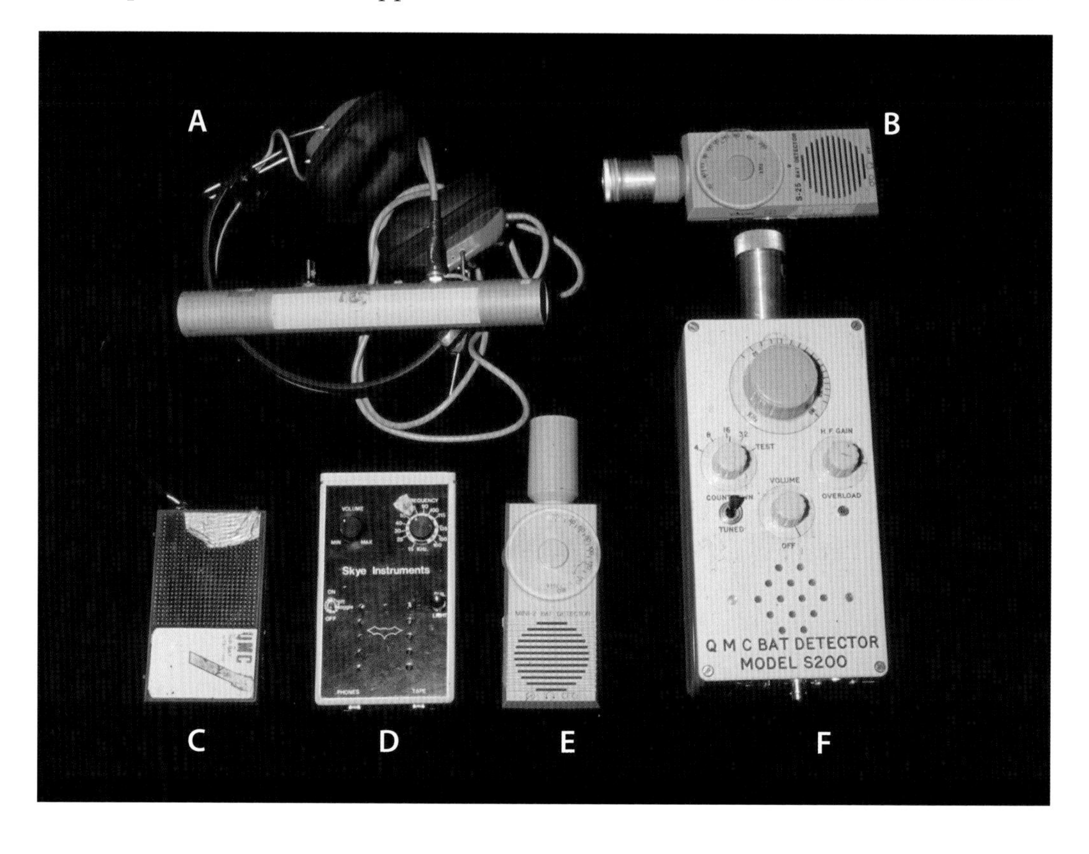

A sampling of the bat detectors that were available in 1982. The leak detector (A) used a crystal microphone tuned to 40 kHz. The other detectors (B through to F) could be tuned to different frequencies. Two of the detectors (B and F) also could record sounds from 5 kHz to over 100 kHz. Detector B could record to 200 kHz, Detector F to 180 kHz.

Biologists as eavesdroppers on bats

The sonic apparatus that Griffin used was an early bat detector. Today bat detectors are essential tools for many bat biologists and there are many kinds at different prices and with different features. Now it is common to use bats' echolocation calls as proxies for what bats do – what species occur where when, and how they use space. Central to this research is the premise that it is possible to identify which species of bat produced the echolocation calls. Observers can also use feeding buzzes to find foraging bats, or playback presentations to document the responses of bats to the calls of others, or to lure bats to nets or traps or to potential roosts. Monitoring echolocation calls allows us to assess possible impacts of habitat changes on bats. This can set the stage for mitigation and making informed choices about how to minimize the impact of developments on these animals.

In the 1950s, the equipment Griffin needed to record bats' echolocation calls in the field filled the back of a pickup truck. Now you can do the job with a laptop loaded with appropriate software and little more than a microphone and some cables. More research and more data about bat calls demonstrate that not all bat detectors are created equal: some are much more sensitive than others. Variation in the signals you have recorded could reflect what the bat said, and/or changes in its position relative to the microphone – microphone quality is critical to determining just what a bat said. You can minimize this problem by deploying an array of several microphones, allowing you to plot an echolocating bat's flight path. However, this process requires considerable, sophisticated analyses.

Plotting flight paths and documenting changes in echolocation calls reveal the dynamic nature of bat behaviour. First, researchers must distinguish variation arising from the setting and the equipment so they can recognize bats' intentional changes in call design. In this way, an observer can document how one bat responds to the presence and calls of another. Observers also can document the behaviour of individual bats, quickly learning that bats are aware of, and respond to, others within earshot. Echolocation signals can serve in a communication role. These findings apply to many different species, including those that eat mainly insects and phyllostomids that mainly consume fruit (e.g. Seba's Short-tailed Fruit Bat).[30] But how do the echolocation calls of one bat interfere with (or jam) the echolocation behaviour of another? Bats have been said to avoid jamming by changing their signals when flying within earshot of others. But it is not convincing to infer jamming avoidance behaviour from changes in calls without individual context.

The dynamic nature of bat echolocation has implications for those trying to identify a species from its echolocation calls. Echolocation calls are not like bird songs, which are usually used for communication by male birds to advertise their presence to other males and to females. Echolocation calls do function in communication but, to date, there is little evidence of bats using these particular signals to advertise their presence to conspecifics, whether male or female. Bats' social calls tend to be longer in duration, limiting their utility in echolocation because the outgoing calls deafen the bat to returning echoes. Still, the echolocation calls of a flying bat have evolved to allow it to gather information about the surroundings.

How can a bat continue to echolocate when flying within earshot of many other bats? Is a bat confused (jammed) by overlapping signals? Well, the answer is probably not. Sometimes Brazilian Free-tailed Bats and European Free-tailed Bats flying close to a conspecific change the frequencies dominating their echolocation calls. Remember, echolocating bats listen specifically for echoes of their own voices, perhaps making them less vulnerable to jamming. The situation will depend upon the scale. Interference could be minimal when there are two bats involved, more challenging if there are 20, and even more when there are hundreds of other bats.

Insect prey

For an echolocating bat,[31] an insect flying in a straight line may be a simple challenge because, in a sonar context, it is a hard object on a soft background. Trains of echolocation calls provide the bat with updates about the insect's position, speed and course. To observe this action, at night go and sit under a light which attracts insects, a good place for bats to hunt. In such a place it can sometimes be easy to see bats approaching, closing with and attacking insects such as moths. This silent entertainment is vastly enriched if you emulate Griffin and use a bat detector to eavesdrop on the bats' echolocation calls. Now you can associate the bats' calls with their flight behaviour and attacks on prey. Hearing a feeding buzz as output from a bat detector is always exciting.

Two Tissue moths hibernating in a mine tunnel also used by hibernating bats. These moths could hear the echolocation calls of the bats with whom they share winter quarters. This leaves open the question of how the moths get in and out without being detected and eaten by the bats.

In this setting, focus on medium-sized moths, and follow the flight path of one around the light. When bats are hunting, it usually is not long before a bat attacks the moth. If you look carefully, you will see some moths taking evasive action such as switching from straight line to zigzagging flight, or just diving for the ground. If you keep score, you usually find that bats miss in about half of the attacks when the moth took evasive action. You are witnessing one disadvantage of echolocation, namely information leakage.

Many species of moth have bat-detecting ears. Ears usually come in pairs allowing the moth to localize an approaching bat from one side or the other. The moth may hear high-intensity bat echolocation calls at 40 m, well before the bat detects the moth. Now the moth just turns and flies away without appearing in the bat's perceptual horizon. Other insects also have bat-detecting ears, including lacewings, crickets, and beetles. Astonishingly, some praying mantises, which only have single ears, can change to erratic flight when they hear a bat bearing down on them.

Brown Long-eared Bats (see pages 138–9) roosting in the peak of a roof regularly bite the wings from their insect prey and drop them. This provides a record of what they have been eating. The orange and black wings in the photograph below are from tortoiseshell butterflies. *JR*

Western Barbastelles (see page 2) use a quiet echolocation mode when hunting some moths that have bat-detecting ears.[32] Their echolocation calls are 10 to 100 times weaker than those of other bats that hunt flying insects. This means that the moths would not hear the bats until it's too late. The cost to the bat is that it would only detect a moth at a distance of 2.2–5 m, compared to as far as 8.7–15 m for bats using stronger calls. The research involved in discovering this not only documented what the Barbastelle said but also identified the moths it ate.

Insects thwart predatory bats in other ways. Scales cover the wings and bodies of moths and butterflies.[33] These scales absorb echolocation calls, reducing the distance at which the bat can detect the moth, and so the threat to the insect. Scales also can protect moths and butterflies from sticky spiderwebs. The moth flies into a web and bounces off, just leaving a few scales behind.

Tiger moths (see page 19) have escalated the bat–moth arms race.[34] When they hear the intense calls of an attacking bat, some species of tiger moth respond by producing trains of clicks which often result in bats aborting their attacks. In other cases, moths' clicks jam bats' echolocation, perhaps by interfering with bats' information processing. At Pinery Provincial Park in Ontario, Canada, researchers caught Dogbane Tiger Moths and, with a pin, disabled the moths' noisemakers. When these muted moths flew off, they were quickly attacked and captured by hunting Eastern Red Bats. But the bats just as quickly dropped bad-tasting moths. Tiger moth caterpillars eat plants (such as dogbane or milkweed) from which they obtain foul-tasting/toxic chemicals. Hence, the moths' clicks could also be signalling to the bat that they have an unpleasant taste. Many tiger moths use distinct warning colouration to signal their bad taste to visually hunting predators such as birds.

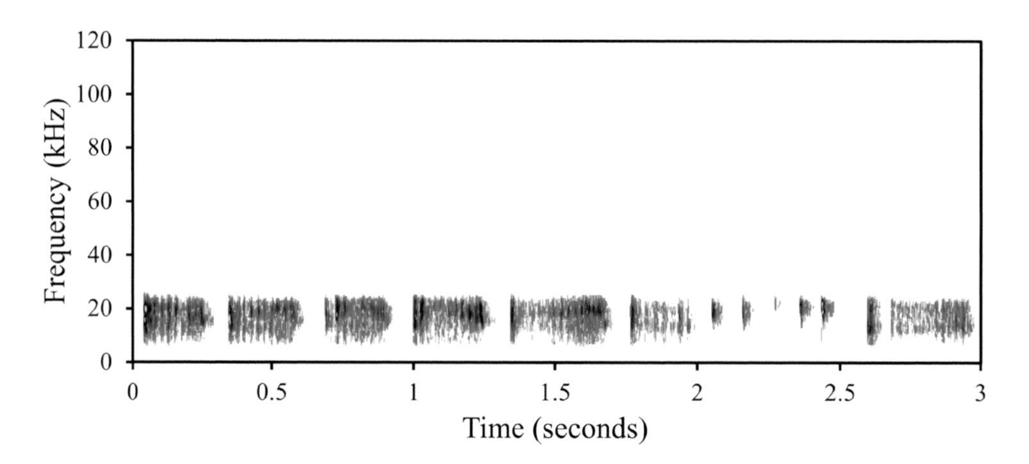

A series of social calls produced by an aggravated Big Brown Bat known as Buzz. Note that most of what the bat said is below 20 kHz, readily audible to a human listener. The calls are not tonal, but broadband, and they sound harsh.

Bat communication

All laryngeally echolocating bats produce both social and echolocation calls. While echolocation calls can have a social function, the reverse is not true (see below). Communication requires a signaller and a signal receiver. Biologists speaking about bat acoustic signals frequently distinguish between echolocation and social calls. The central difference between the two is revealed by changes in the behaviour of the receiver. Echolocation calls dominate if you are listening to a bat hunting around a light. When another bat arrives on the scene you can expect a mixture of echolocation and social calls, especially if one bat chases the other away. Echolocation signals tend to be tonal and broadband, while aggressive social calls are harsh sounding.

Spix's Disk-winged Bats roost in furled leaves and usually must switch roosts almost daily. These bats use calls of others to find the leaf roost in use on any day (see page 135).

We know relatively little about the vocal repertoires of bats. Bats often have a lot to say, but much of it is beyond our hearing. As usual in mammals and birds, tonal signals (such as whistles) are more pleasant to the ear than harsh broadband hissing sounds.

Many bats 'sing', for example male Greater Sac-winged Bats. Bats also produce higher call repetition rates, known as buzzes. Irritation buzzes warn other bats away from a preferred roost site or food. Florida Bonneted Bats, for instance, mix social and echolocation calls, often to effect social integration. Similarly, Brazilian Free-tailed Bats often use a combination of social and echolocation calls. The incidence of songs around roosts is highest during the season when these animals are mating. We need more information about the social lives of bats to better understand the importance of acoustic signals, whether echolocation or social.

All species of bats can probably produce 'distress calls', perhaps the equivalent of a 'mayday' call from a human. When associated with a predator, one theory suggests that distress calls attract other bats, perhaps to swarm the predator that has caught the calling animal. Alternatively, distress calls could warn other bats away. Distress calls tend to be harsh sounding, lower in frequency and longer in duration than echolocation. Distress calls are also the basis of some lures intended to attract bats. The calls of a bat caught in a net or bat trap may attract others, a ploy often used to catch more bats.

Young bats and their mothers communicate with vocalizations. Signals from young help a mother bat to find and reunite with her offspring. A mother bat also depends upon the odour of her pup(s), partly derived from the smell of her milk, as well as the odour of the pup itself. Some bats have well-developed glands on each side of the muzzle. Bats can use secretions from these pararhinal glands to mark the roosting surface, as well as dependant young. While mother bats are very picky about which young can drink their milk, other bats' pups readily 'steal' any milk they can get. From a genetic standpoint, a mother wants to invest only in her own offspring and may vigorously dissuade would-be milk thieves.

Bats use special calls to attract others and guide them to a roost. Well-studied examples are Pallid Bats from Arizona and Spix's Disk-winged Bats (see page 135) from Costa Rica.[35] To explore such interactions, biologists used playback presentations of bat calls, and odour choice experiments. Bat communication often involves combinations of signals in different modalities: acoustic, visual, tactile, chemical and body language. Male Greater Sac-winged Bat (see page 155) displays combine visual (flight), acoustic (songs, echolocation calls) and olfactory components. Bats' vocalizations cover a broader range of frequencies than humans', from ~5 kHz to over 200 kHz (in general humans hear from about 40 Hz to 20,000 Hz). At least some of the time, receivers can be eavesdroppers, so communication covers a variety of situations and players. We expect to hear much more about detailed communications in social networks of bats in the future.

Mexican Funnel-eared Bat, a small insectivorous species that usually roosts in caves

Davis's Round-eared Bat, an example of a phyllostomid that appears to eat both fruit and insects

Air traffic control

We often observe high densities of flying bats in situations where resources, or access to them, is limited. Such aggregations also occur around roosts and areas with high densities of prey. It is tempting to suppose that bats have a system of air traffic control that involves monitoring echolocation calls. This could minimize the chances of mid-air collisions which would be disruptive or damaging to the participants. At night during migration, large numbers of nectar-feeding bats (Lesser Long-nosed Bats and Mexican Long-tongued Bats) converge on hummingbird feeders at the Southwest Research Station in southeastern Arizona. We have not observed collisions there, even when scores of bats fly in the same airspace, jockeying for position at the feeders.

At a more local level, bats may slightly modify their echolocation calls to add a social component. They do this by broadening the bandwidth of the calls and 'honking'. Greater Bulldog Bats hunt for fish swimming close enough to the surface that they break the water. The honks they emit allow them to avoid a mid-air collision.

To learn more about bat communication behaviour we could use playback presentations of echolocation calls and see if they affect bats' flight behaviour in crowded airspaces. As shown, Common Vampire Bats flying out of a tunnel in Belize fly higher than those flying back in, which may be effective air traffic control.

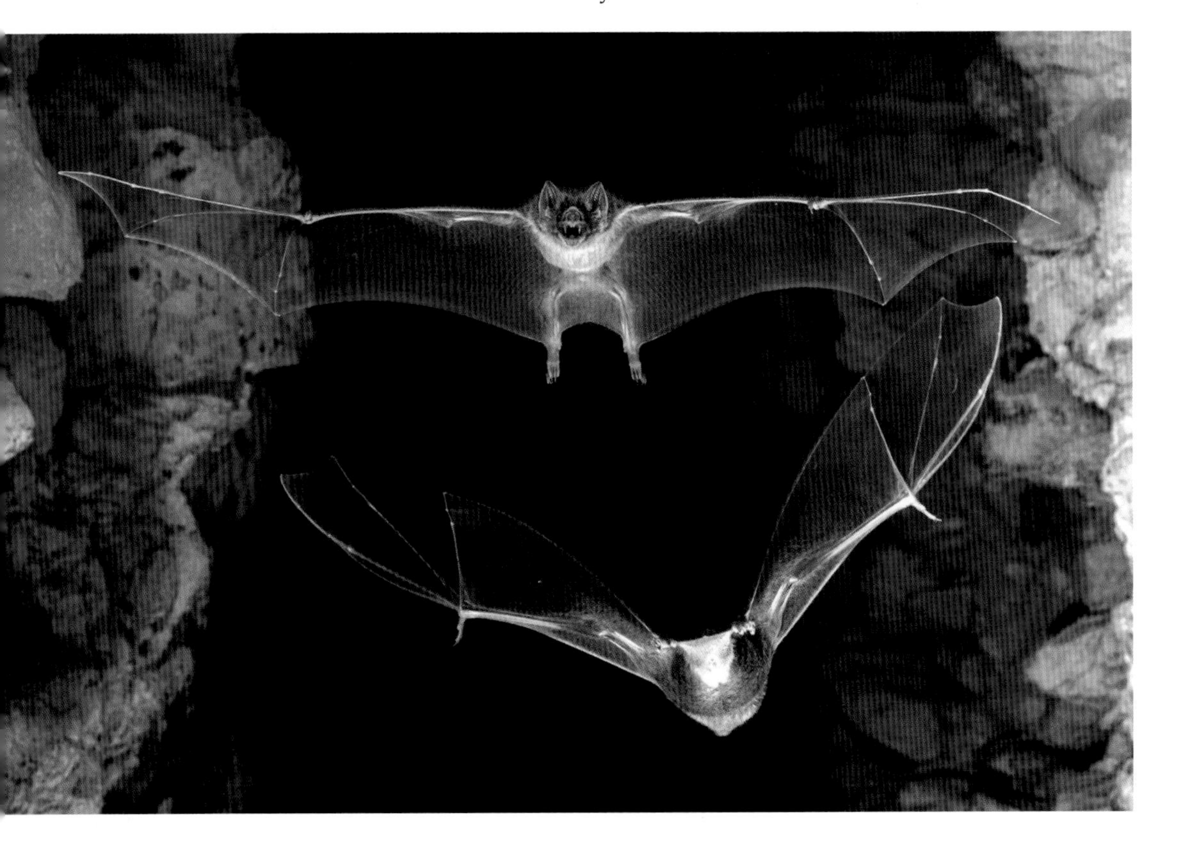

Two Common Vampire Bats flying in the same tunnel at the same time. Note that the emerging bat (face on) flies high, while the returning bat flies low.

Three successive views of a Mexican Long-tongued Bat approaching and then turning away from Agave blooms which it visited to obtain nectar and pollen, in south-eastern Arizona.

Echolocation and foraging

A pregnant female Fringe-lipped Bat (35g)

The discovery that some bats exploited other animals' courtship sounds was considered remarkable because 'everyone' presumed that the bats relied on echolocation. In an early study in Panama, Robert Barclay, Merlin Tuttle and others showed that Fringe-lipped Bats located túngara frogs by listening for the calls of males.[36] Here the bats also approached a tape recorder playing the frogs' calls. These bats used calls to distinguish good-tasting frogs from bad-tasting toads. The bats could also be trained to come to the sound of a Bob Marley song being played.

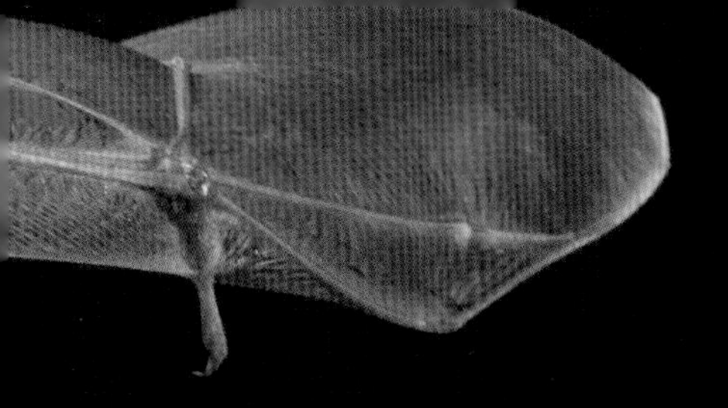

Other recordings revealed that Fringe-lipped Bats were echolocating as they approached a singing frog. Túngara frogs cannot hear the bats' echolocation calls, so there is no reason for the bat to stop echolocating when it is attacking one. In India, Greater False Vampire Bats hunt frogs as well as mice. At first it appeared that these bats stopped echolocating when hunting – this made sense because mice have good high-frequency hearing and might have detected the approaching bats. Further research demonstrated that Greater False Vampire Bats echolocated when hunting or flying in unfamiliar surroundings, but not when they knew the terrain.[37] Both Fringe-lipped Bats and Greater False Vampire Bats are examples of what Griffin would have called 'whispering' bats.

In India, a Lesser False Vampire Bat uses the calls of male katydids to locate these prey. Male katydids often aggregate when they stridulate. The bats detected calling aggregations from greater range than single callers, but experiments revealed that the bats took longer to catch a katydid calling from an aggregation than one calling alone because it was harder for them to home in on a single insect.[38] This is another example of a trade-off – for the bats and for the katydids.

A Lesser False Vampire Bat (25g). The two species on these pages produce quiet echolocation calls while using prey-generated sounds to detect and locate prey.

5 What bats eat, part I

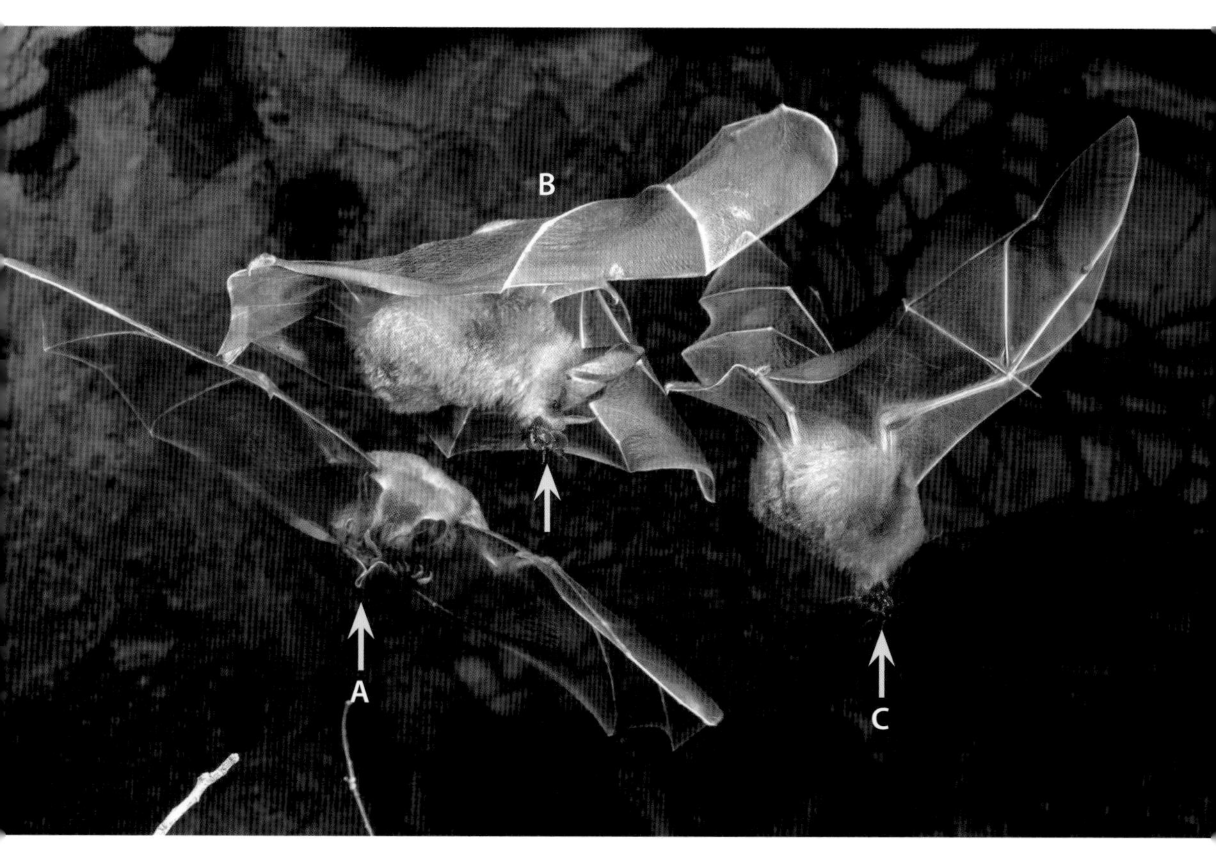

In addition to flight and echolocation, bats are renowned for their diversity of diets. While most of the 1,400 or so living species eat mainly animals, others consume fruit, nectar and pollen. Among animal-eaters, insects prevail as food, but the menu also includes fish, frogs, birds, mammals, other bats and blood (adult vampire bats). With more details about what bats actually eat, we can see that some species classified as 'fruit' eaters, also consume flowers (nectar and pollen) and insects.[39] Some so-called fishing bats actually eat insects and even other bats. Among bats in general, the three species of blood-feeding vampires are the best fit to our traditional classification of bats by diet – the adults appear to feed only on blood.

Three successive views (A, B, C) of a Golden Bat (35 g) with a captured beetle (arrows). In A, the bat has caught the beetle in the membrane between its hind legs and is transferring it to its mouth. In B, the bat has the beetle in its mouth, and in C flies away with it. SLF and MBF took the photograph at the entrance to a cave in Belize when the bat triggered flashes three times while the camera shutter was open.

A Black Mastiff Bat (opposite), a 30 g insectivore from Central and South America. The gular (sebaceous) gland under the bat's chin identifies it as a mature male (see also page 12).

Bats can eat a great deal of food every night. Imagine your favourite food. Now think of how much you weigh. Then imagine eating half your body mass in your favourite food even once, let alone every night. Bats typically eat this much. Lactating females may eat more than their body weight in food every night. Energy and size help to put bats' diets in perspective. First, bats are high-energy animals. Heart rate is a good indicator of what high energy means. A flying bat's heartbeat may be 1,200 times a minute, reflecting high levels of energy consumption. The same individual's (e.g. a Little Brown Myotis) heartbeat rate drops to around 300 beats a minute when it lands. Such levels of energy consumption coincide with high food intake. Nursing pups eat their own weight in milk every day, putting lactating female bats under considerable pressure.

Learning how much a bat consumes

Using radio-tracking, researchers determined how much time Eastern Red Bats and Hoary Bats (see pages 60 and 61) spent foraging in Pinery Provincial Park (Ontario, Canada).[40] Both species hunted among concentrations of insects around streetlights. By watching them hunting, researchers determined how many insects each bat ate. They also used a Doppler radar to measure flight speeds. Then, with measurements of body mass and wing elements, they calculated how much energy hunting bats consumed.

Two assistants, Ben and Eleanor, use pizzas to demonstrate the relative amounts of food male and female bats need to eat. For Ben, it's half his body weight in pizzas and for Eleanor, her entire weight in pizzas.

Hoary Bats are more than twice as big as Eastern Red Bats. They usually ate larger moths (weighing on average 44 mg, versus 30 mg). Both species succeeded in 40–60% of their attacks on insects (mainly moths) and individuals of both species roosted about 500 m from their foraging areas. On average, Hoary Bats foraged for 222 min per night, Eastern Red Bats for 127 min. On average, Hoary Bats attacked a moth every 17.45 seconds, Eastern Red Bats every 20 seconds. The estimated flight costs were 1.14 kJ (which required them to eat 19 moths weighing 30 mg) for Eastern Red Bats, 5 kJ for Hoary Bats (or 28 moths of 44 mg).

Hunting in concentrations of insects permitted both species to meet their costs of operation (commuting, hunting and, for females, lactating) in the time they spent foraging. Working with Hoary Bats in Manitoba, another researcher found that Hoary Bats attacked a prey item about once every 62 seconds, flying and foraging all night to break even.

Researchers working in the Yucatan Peninsula in Mexico employed radio-tracking to study Black Mastiff Bats (see page 74).[41] Six of these 30–35 g bats each foraged for 22–63 minutes a night. Bats returning to the roost had distended stomachs. One female produced 0.9 grams of faeces after foraging for 16 minutes! She and the other bats ate mainly beetles, apparently caught within about 2 km of their roost. In other words, insectivorous bats eat large numbers of insects and do so as rapidly as possible.

Second, size matters when it comes to diet, specifically the size of the bat and the size of the prey. Felix the cat came home and proudly presented JR with a captured bird, a 20 g Great Tit. There was a Northern Bat (10 g) attached to the bird's neck by a powerful bite. The bird was dead but still warm. The bat was alive and stubbornly refused to relinquish his prey. The bat had caught the bird, killed it and might have eaten it had the cat not intervened. Normally, Northern Bats eat mainly insects, each smaller than 1 g in the main. Had the bat mistaken the bird for something else?

One of us (MBF) watched a 4 g California Myotis hunting insects around a streetlight in British Columbia. The bat usually attacked, caught and ate smaller moths and beetles. But then it decided to repeatedly attack a flying cecropia moth (weighing in at 1 g and with an 18-cm wingspan), a large undertaking for a small (22-cm wingspan) bat. The bat persisted in its attack and eventually both moth and bat ended up on the ground. Each flew away, apparently none the worse for their tussle.

The Northern Bat and the California Myotis demonstrate that bats sometimes bite off more than they can chew. MBF watched a Large Slit-faced Bat (30 g; see page 79) chase, catch and kill a flying Sundevall's Roundleaf Bat (10 g). The Large Slit-faced Bat took its prey to a roost and spent over 90 min of chewing to ingest the meal. Hard to tell if the meal was worth the effort. Many bats that catch larger prey take it to a roost to eat it.

So, do bats choose their prey carefully or just take what they can get? Returning to watching hunting bats, when you look closely and are patient, you will see that they seem to attack different kinds of insects. If you gently toss small (1-cm diameter) balls of plasticine into the air, Eastern Red Bats often will chase and grab them. This works on the first night, but after the bats have learned to pay more attention to the information they get from echolocation, they stick to insects.

To further explore some details of bat foraging, bat biologists worked over the still waters of six lakes in southern British Columbia (Canada).[42] They deployed targets in

the afternoon before the bats (Little Brown Myotis and Yuma Myotis) arrived to hunt. Three types of targets were suspended 3–10 cm above the water, about 5 m from shore. These were either edible prey in the form of small moths (12 mm long) or large moths (32 mm) or inedible prey including meal-worms (which were too large for the bats), leaves the size of small moths or those the size of large moths. In some experiments the targets were moving, in others they were stationary. The bats always attacked moving targets much more often than stationary ones. They consistently attacked small insects more often than large ones.

What bats catch and eat usually reflects what is available where and when they are hunting. This means that the actual diet of a bat (species or individual) can vary considerably, making it difficult to categorize bats by their diets.

These were either edible prey in the form of small moths (12 mm long) or large moths (32 mm) or inedible prey including meal-worms (which were too large for the bats), leaves the size of small moths or those the size of large moths. In some experiments the targets were moving, in others they were stationary. The bats always attacked moving targets much more often than stationary ones. They consistently attacked small insects more often than large ones.

What bats catch and eat usually reflects what is available where and when they are hunting. This means that the actual diet of a bat (species or individual) can vary considerably, making it difficult to categorize bats by their diets.

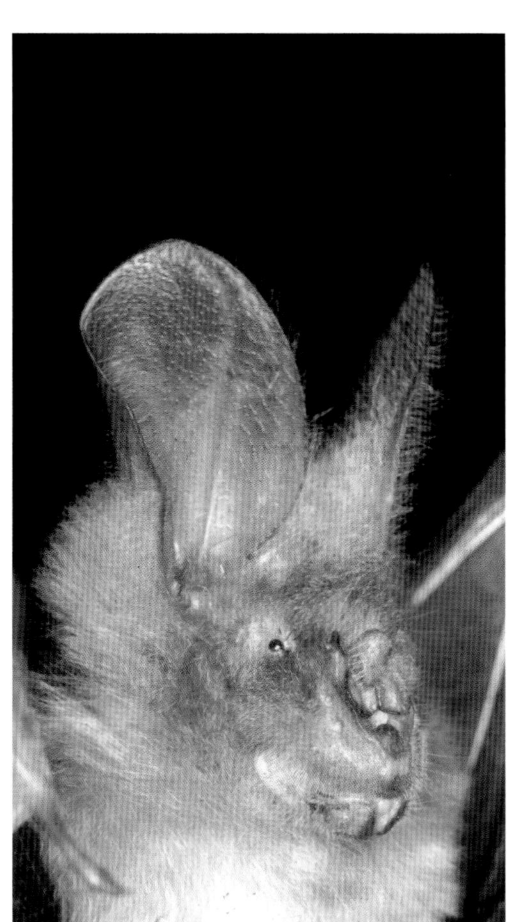

A Large Slit-faced Bat from Mana Pools National Park in Zimbabwe. This species is widespread in Africa, but little studied.

Black Mastiff Bats (opposite) come in black and red. Two colour varieties are common in several other species and do not indicate taxonomic differences. *JR*

Some bats eat birds

Working in the south of Spain, researchers found that Greater Noctules (46 g) preyed on migrating songbirds ranging up to 23 g in body size.[43] In Japan, Bird-like Noctules (46 g) also eat birds, as do Greater Evening Bats (49 g) in India and in China. These bat species all appear to catch and eat smaller birds on the wing. The other bats that may eat birds have broader wings and may more often take their prey to a roost to consume it. These include Greater False Vampire Bats (39 g), Lesser False Vampire Bats (25 g) and even some larger species (Big-eared Woolly Bat – 75 g, see page 82; Ghost Bat – 130 g; and Spectral Bat – 150 g, see page 3). Spectral bats take the nest contents of Groove-billed Ani – incubating adult, young, eggs and all.

Egyptian Slit-faced Bat, a 10 g insectivore sometimes taken by Large Slit-faced Bats (see page 82).

Three species of noctules range in size from 12 g (Lesser Noctule/Leisler's Bat, above) to 28 g (Common Noctule, opposite top) to 46 g (Greater Noctule, opposite bottom). The largest species eats migrating birds as well as a variety of insects. *JR*

A Big-eared Woolly Bat is another large, versatile predator.

Versatility

In 1979 in Zimbabwe we heard about frog-eating bats at a site in Mana Pools National Park. A group of up to 60 Large Slit-faced Bats roosted on the ceiling of a 5-m tall cement structure that had supported a water tank. The ground below the roost was littered with the remains of their prey. At night the tower resounded with the sounds of feeding bats – lip-smacking, loud chewing and bone crunching.

We went back to the site several times over the next 10 years. From discarded prey parts we learned that these 30–40-g bats ate frogs, fish, birds, other bats, as well as a variety of arthropods, from sunspiders to insects, including large hawkmoths and cicadas. This was exciting because this was the first African bat species shown to eat other vertebrates. Radio transmitters on the bats revealed that they also roosted in hollow acacia trees or in disused military bunkers. Each night radio-tagged bats hunted along the shores of the Zambezi River, but their night/feeding roosts were scattered among the buildings constituting the park headquarters.

Although we spent many nights walking and following these tagged Large Slit-faced Bats, we never saw one catching fish. They took frogs from along the shore and from the water's surface. They appeared to take birds from their night perches, and hunted and caught a variety of other bats, including 10 g Egyptian Slit-faced Bats. We were not able to show that the predatory bats eavesdropped on the echolocation calls of their bat prey – but this remains a possibility.

Large Slit-faced Bats are versatile hunters. When handling captured individuals, we quickly learned that they have a powerful bite. We also observed that they killed other bats with a powerful bite to the face, reminiscent of cats. Our fieldwork on these bats involved several challenges, not least of which was walking about at night following bats and avoiding some of the other much larger mammals that occurred there (e.g. elephants, lions, hippos, buffalo and rhino) adding to the variety of nocturnal thrills.

Around the tropics, big-eared species of bats weighing 30–50 g eat a variety of animals often captured from surfaces such as the ground, tree branches or foliage. These bats use the sounds of prey to locate their targets. The menu can include other bats, birds and various reptiles and amphibians as well as a variety of large arthropods.[44] The list of such bats includes Yellow-winged Bats and Heart-nosed Bats (Africa), Greater and Lesser False Vampire Bats (India, Southeast Asia), and a variety of leaf-nosed bats from the New World, including Golden Bats and Fringe-lipped Bats. Some other predatory bats are much larger. In the New World, Big-eared Woolly Bats can weigh more than 75 g. Spectral Bats (see page 3) weigh 150–200 g, as do Australian Ghost Bats. Larger predators may take larger prey, perhaps leaving smaller targets for other species.

What insects do bats eat?

The high energy costs of flight means that bats have to ingest large quantities of food (for their size). Their energy demands sometimes translate into varied diets. Therefore, saying that a bat is insectivorous is a generalization. Insects are very diverse, in terms of numbers of species, sizes and life histories. Bats tend to rapidly process their food, from handling to chewing, swallowing and digesting. Anyone who has examined closely the stomach contents or droppings of an insectivorous bat will appreciate that the macerated mass could be the remains of many individuals and/or species of insect prey.

A male Spongy Moth and its DNA barcode.

How can you identify food remains from bat gut contents or droppings? DNA barcode analysis provides one handy solution.[45] This process depends on the CO1 mitochondrial gene which encodes cytochrome C oxidase subunit 1. Biologists use polymerase chain reactions (PCR – which might now be familiar to many more of us following self-testing during the Covid-19 pandemic!) to produce sequences of CO1

mitochondrial genes obtained from fragments of insects in bat droppings. This is reminiscent of using the barcode on an item you purchase at a store to determine what it is and what it costs. The CO1 gene can be used to identify the insect species bats have eaten – provided there is a record of the relevant CO1 gene available for comparison. At a more general level, this method can tell you the number of different species of insects in the sample without knowing species' names. Using this approach, biologists collected bat droppings from sites across Canada and showed that Little Brown Myotis ate over 900 species of insects. This allowed bat biologists to go beyond the 'eats beetles' or 'eats moths' to being very specific about what's on a bat's menu.

The high costs of flight may oblige bats to be selective. A bat may choose to search for concentrations of prey, such as insects around a light, to minimize the time between encounters with targets. Then it takes the largest prey it can handle. This raises questions about just what the California Myotis or Northern Bat were 'thinking' when they attacked very large (for them) prey. When the insects are in a mating swarm, the bat could fill up on one species. Around a light there may be several species, and so the diet will reflect this. For the interested biologist, DNA barcode analysis offers a way to explore the topic in more detail. DNA barcode analysis has even revealed that one pregnant Greater Evening Bat ate a Grasshopper Warbler (13 g), supporting the suggestion that this species may eat birds.

Specialized hunting

Researchers in Japan demonstrated that Japanese House Bats (6 g) echolocate when hunting small insects. By reconstructing bats' flight paths, they showed that the bats often started tracking a second insect before they had attacked and caught the first. This revealed how these bats exploited swarms of insects. When hunting, they planned ahead, minimizing the time between captures.[46] This strategy surely increased their food intake and likely shortened their overall foraging times.[47]

In Israel, scientists used computer simulations to demonstrate that echolocating bats could detect swarms of insects. Indeed, they detected swarms at greater distances than individual flying insects.[48] This is another way in which bats' behaviour increased their rate of food intake and could shorten feeding time. Both studies illustrate an advantage of echolocation when hunting flying insects and support the argument in favour of the immediate ancestors of bats also being able to echolocate.

Researchers studying the echolocation and foraging behaviour of captive Big Brown Bats revealed that individual bats accumulated evidence across sequences of echoes. These bats used this information to predict the flight trajectory of a target. This behaviour permitted them to increase their foraging efficiency under uncertain conditions.

Lesser Moustached Bat. Like other bats that hunt over water (such as Proboscis Bat, pages 128–9; Little Brown Myotis, page 89), they do not have enlarged hind feet. Around the world species of bats with enlarged hind feet are known as trawlers (e.g., Rickett's Big-footed Bat). *JR*

Daubenton's Bat hunting over water. *JR*

Trawling

Many species of insect fly over water such as lakes, rivers and ponds. No surprise, then, that many species of bat hunt over water too. Perhaps bats have an advantage in these situations as the insects have nowhere to hide and occur in a smaller air space. Almost everywhere in the world trawling bats, such as Daubenton's Bat (8 g), use their feet to take prey from the water's surface.

Trawling bats fly very close to the water, as little as 15–20 cm above it, and shorten their wingbeats to avoid dipping them in the water. They direct their echolocation beam straight ahead and parallel to the water's surface.[49] This ensures that reflected sound will bounce away from the bat, while echoes from objects on the surface come to the bat. Flying higher above the water and directing the echolocation beam at an angle to the water would bombard the bat with echoes from water, which could mask echoes from insects.

Daubenton's Bats tend to congregate on the leeward side of lakes or over small streams and ponds because trawling is less effective over rippled water. When there is no calm water, Daubenton's Bats will go and hunt in nearby forest. These bats also fly higher over water with small ripples and do not attempt to hunt there. Why not try? What makes the forest a better alternative than a slightly rippled surface? Well, a microphone placed near the small ripples (2–3 cm) on the water surface detected many short clicks. These appeared to resemble echoes from small insects, perhaps making it more difficult for a bat to detect and track a flying insect. Daubenton's Bats detect echoes from small insects at about a metre, so there may not be enough time for the bat to discriminate echoes from ripples from those of insects. An abundance of false echoes seems to preclude trawling over rippled water.

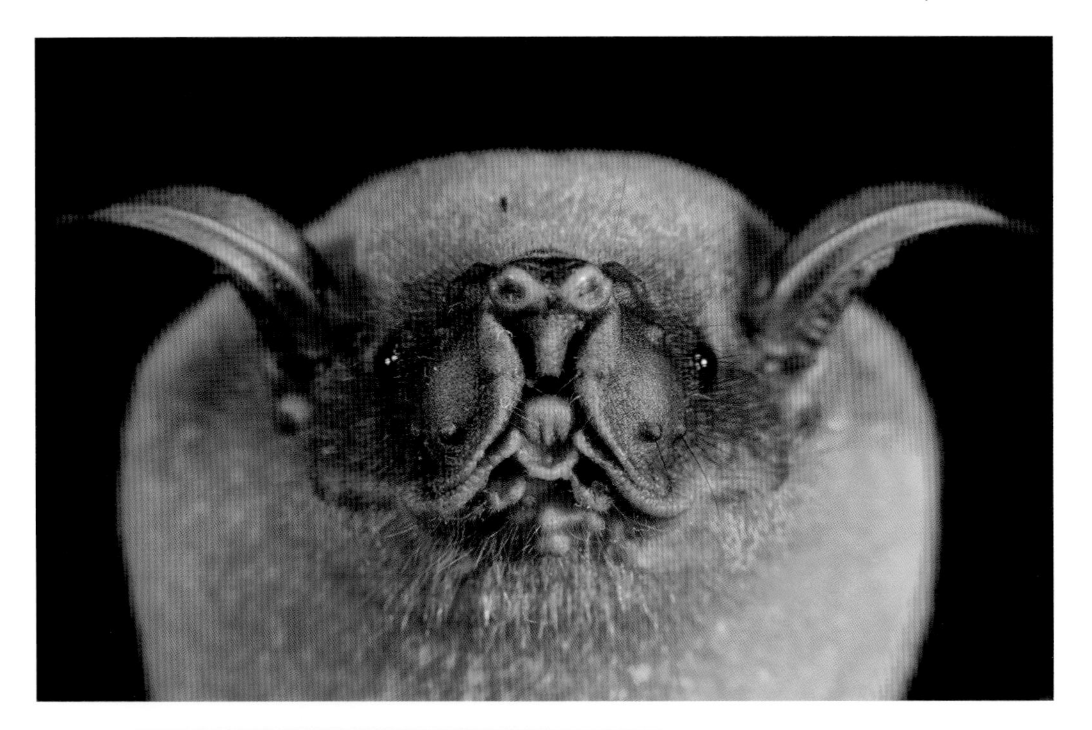

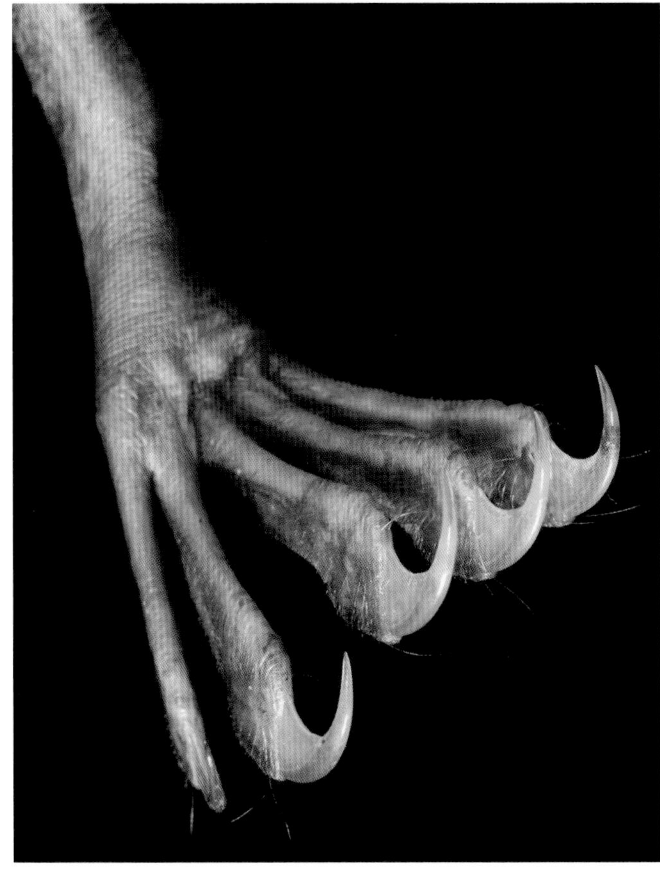

A portrait of a Greater Bulldog Bat, which has distinctive hind feet and claws. The hind foot is 30 mm long.

Flying near the ground or just above the water, bats can benefit from the so-called ground effect, saving some of the aerodynamic cost of flight (see also page 32). Ground and water surfaces act as an aerodynamic mirror, increasing air pressure under the wing. This provides some extra lift and reduces the cost of flying. Working in a wind tunnel, researchers accurately measured the ground effect. By staying low over the water, a Daubenton's Bat may save as much as 30% of the cost of flight. This exceeds

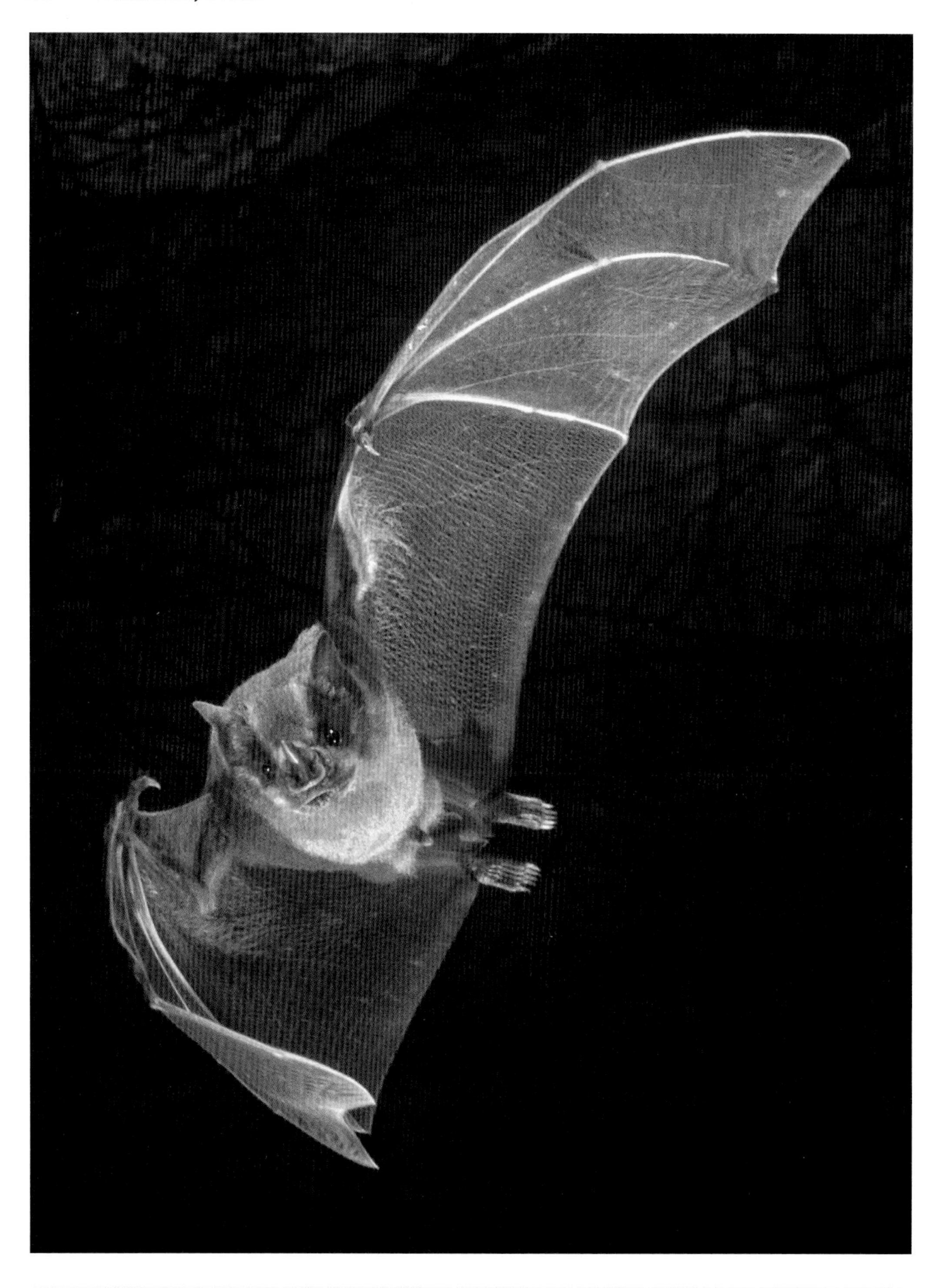

The toes of flying bats may be held close together as in this Jamaican Fruit Bat (above) or well apart as with the Greater Bulldog Bat (right).

the amount predicted from theory. The ground effect makes trawling an even more efficient way for bats to hunt.

Bats that trawl may catch other prey as well. In 1975, researchers found distinctive fishbones (otoliths) in the digestive tracts of Daubenton's Bat, revealing that this species did not limit itself to insects when hunting over water. By 2000, there was considerable evidence that fish eating was widespread in trawling bats. Other small (< 10 g) species also turned up as fish-eaters. Greater Bulldog Bats (55–70 g) may be the best known 'fishers' of the trawling bats. These bats hunt over water where they catch fish, and over sand beaches where they also take hermit crabs and even other bats.[50] Fishing Myotis are about half the size of Greater Bulldog Bats and some hunt over salt water, taking crustaceans as well as small fish.

It is interesting to contemplate the relationship between bats drinking and trawling in bats. Little Brown Myotis are similar in behaviour and morphology to Daubenton's Bats. They often hunt over water and appear, at least, to take insects from the surface; to date there is no evidence of them catching fish. We have often watched one of these bats exploiting moths that dived in response to the echolocation calls of an approaching bat. Having dived, the moth ended up on the water's surface. Sometimes we think the bat circled back and took the moth from the water. A flying insect can move in three dimensions (up, down, sideways) while the insect on water can move in only one plane – hence the bat would have turned a problem with three dimensions into one with just two.

The toenails of some bats are very sharp. They use them to hang from quite smooth surfaces (but see page 134 for bats that also have suckers on their ankles). Grooming bats use their toenails to comb their fur. Many species typically fly with their toes held tight together. But Greater Bulldog Bats, which catch fish with their sharp claws, may fly with their toes wide apart.

Diets of bats

Many people, even bat biologists, seem compelled to categorize things. So, we speak of insectivorous bats, or frugivorous bats, knowing that there is a huge variety of insects, and that fruit eaters may also consume leaves as well as flowers. If you think about what you have eaten in the last month and had to categorize yourself by diet, where would you fall? If you were on a strict diet, it is likely that you can be specific about what you ate during the last month, not just what but when and how much. How many vegetables does it take to move you out of the meat-eating category? How much does your diet reflect the types of food that are available?

Research on the microbiota (bacteria and fungi in the gut) of bats has revealed that only the blood-feeders (sanguinivores) are accurately classified by diet. Some samples from other species – whether classified as insectivores, frugivores, nectarivores, piscivores or carnivores – invariably include evidence of other food. At one level this means insect-eaters also take spiders or scorpions, fruit-eaters also eat insects or other small animals. Furthermore, some of the nectar- and pollen-feeders regularly consume insects and other arthropods, while piscivores also take insects and crustaceans or other animals. The problem with 'carnivorous' (literally meat-eaters) is the definition. Some larger (> 50 g) 'fruit' bats such as Leschenault's Rousette also eat fish.[51]

Sometimes a fruit-eating bat may ingest insects on or in the fruit, the same with flower-visitors. More generally, as high-energy animals, bats are apt to eat whatever is available.

A heavily pregnant Fringe-lipped Bat emerging from its roost in a cave.

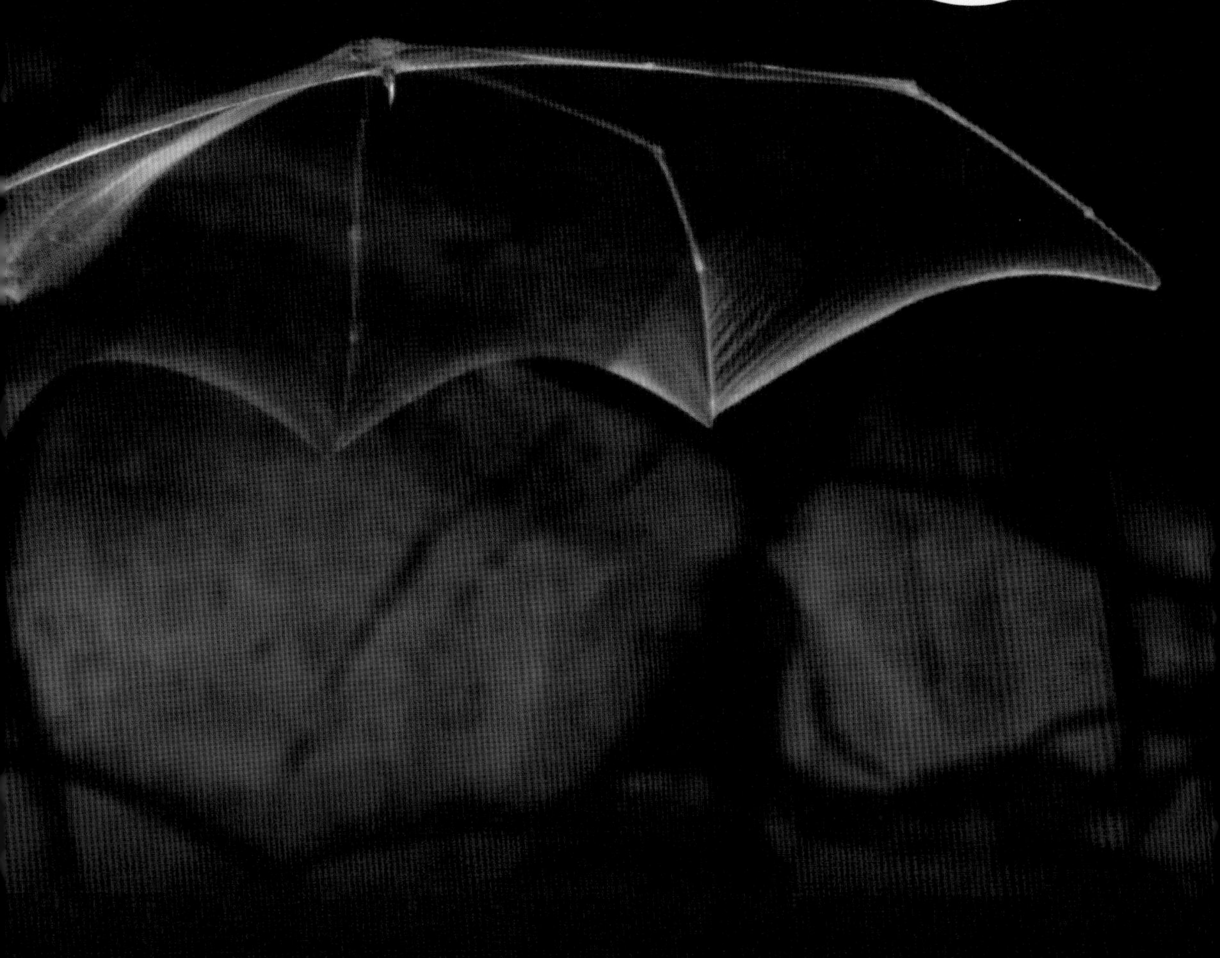

6 What bats eat, part II

This Jamaican Fruit Bat was photographed as it flew through a cave in Cuba. Originally described from Jamaica, the species is widespread in the Caribbean, as well as Central and South America. We may eventually learn that there is more than one species in this complex.

It has been traditional to categorize bats by their diet – as insect-eaters (insectivores), fruit-eaters (e.g. frugivores), nectar- and pollen-feeders (nectarivores), fish-eaters (piscivores), those that eat meat (vertebrates but not fish – carnivores) and blood-feeders (sanguinivores). Pallas's Long-tongued Bat (see also page 102) as often as not eats fruit, nectar and pollen, and insects. The microbiota of bats, their gut flora, reveals more complicated pictures and often suggests that Pallas's Long-tongued Bat is the rule rather than the exception when it comes to its diet.[52] This work involves using a variation on DNA barcode analysis (DNA metabarcoding) to explore the gut faunas of a selection of bats. Data from 25 species from sites in Belize revealed that vampire bats were the only ones

Ridges on the roof of an Egyptian Fruit Bat's mouth (opposite) are used in combination with it's tongue and teeth to mash up ingested fruit. Species such as Jamaican Fruit Bats and other phyllostomids use the same behaviour.

Seba's Short-tailed Fruit Bat

whose diets were accurately described by their classification (as sanguinivores). Other species labelled as insectivores also ate plant material, while some treated as frugivores also ate insects as well as flowers (nectar and pollen). An underlying question is how much a bat's gut flora reflects its evolutionary history, and how far this influences its diet. This is a tantalizing question, and the focus for much current research.

Common Vampire Bats are an interesting example. In Belize, researchers sampled droppings of these bats in two areas (separated by about 8 km), one a smaller forest fragment, the other a more continuous forest. The data showed that even at this level of separation the bats' diets and microbiota differed. These findings have implications for how land-use change can influence the distribution of livestock, and in turn the vampire bats and their microbiota.

Certain species use ridges on their palate in conjunction with their tongues and teeth to extract fruit juice and pulp. By sucking as they use the tongue to squeeze fruit pulp against the ridges, the bats swallow the material they can digest and spit out ejecta, resulting in a pellet of plant fibre (see page 96). Seba's Short-tailed Fruit Bat uses this same approach when eating insects such as cockroaches. Now the ejecta mostly consist of insect exoskeletal remains. This presents a problem to people studying the bat's diet by analysing its droppings: insects remains are absent from the droppings, concealing the make-up of at least some of their diet.

Fruit-eating species

Beyond general orientation and avoiding obstacles, the role that echolocation plays in the foraging behaviour of fruit-eating bats is less clear. Furthermore, there is a sharp distinction between the fruit-eating bats of the Old World (flying foxes and their relatives) and the New World leaf-nosed bats (phyllostomids). Most Old World fruit bats do not echolocate, and those that do use tongue clicks rather than signals produced in the larynx (see page 47). New World leaf-nosed bats use laryngeally produced echolocation signals. Both kinds of fruit bats identify ripe fruits from their odour, and know where to find fruits such as figs. While some fruit-eating bats are generalists, searching far and wide for ripe fruit, others predictably visit productive fruit trees. Seba's Short-tailed Fruit Bats readily find concentrations of fruit deployed by researchers, while Jamaican Fruit Bats virtually never visit such sites, going instead to trees with ripe figs. When fruits are larger, bats often take them to a night roost to consume, dropping inedible parts such as hard seeds. These often sprout under the roost, and in dark roosts they produce ghostly plants in the absence of light. A fruit bat may handle its own weight in fruit every night but ingest only the digestible portions of the food.

In the laboratory, Northern Yellow-shouldered Bats trained to respond to solutions with infusions of banana, quickly located food sources by odour and readily followed odour trails to food sources.[53] These bats use a combination of side-to-side head movements to identify and follow odour gradients and their spatial awareness to locate food. In behavioural trials, bats approached food sources from above and below, according to the situation. A bat's success was influenced by the strength of the odour signal.

A plant that has sprouted from a seed dropped to a cave floor in Cuba

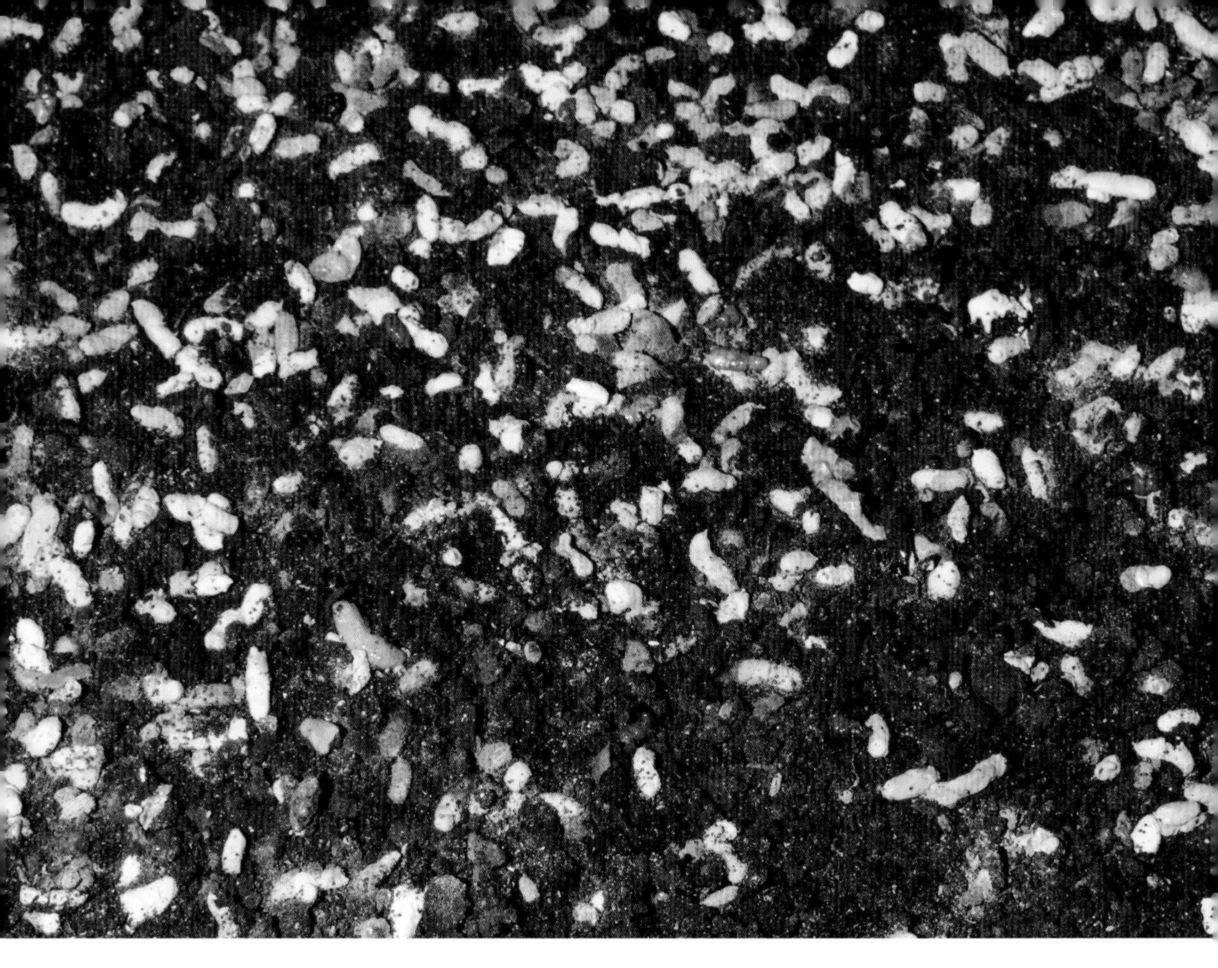

The cave floor under a roost used by Jamaican Fruit Bats is covered with ejected pellets and droppings. The finger provides a scale. In this, and other bat roosts, it is never wise to look up with your mouth open.

Northern Yellow-shouldered Bats: shoulder glands (upper image, arrow) in adult males distinguish them from females (lower image).

Bats and flowers

Bats are the main pollinators for over 500 species of plants, a phenomenon known as chiropterophily.[54] Bat-pollinated flowers tend to be large, white or whitish in colour, open at night, have a strong musty odour, and produce large quantities of nectar. Examples of human foods pollinated by bats include mangoes, guavas, durians and agaves.

Some bat-pollinated flowers, including Shaving-brush Tree *Pseudobombax ellipticum* (above), Devil's-guts Cactus *Selenicereus testudo* (left) and Palmer's Agave *Agave palmeri* (right), the latter approached by a thirsty Mexican Long-tongued Bat.

New World bats specialized for eating nectar and pollen sometimes show conspicuous flight antics (see page 35).[55] Capacity for manoeuvrable flight allows the bats access to flowers that would otherwise be unavailable. Hovering is one aspect of their flight and gives the animal a stable position while drinking nectar.[56] This technique is a fundamental feature of New World leaf-nosed nectar-feeding bats. These bats take different approaches to hovering than the more notorious hummingbirds, but in both cases the cost of hovering is the same per unit of body mass.[57]

Like hummingbirds, the wings of hovering nectar-feeding bats are palm down during the downstroke and palm up on the upstroke. In hummingbirds, wing shapes are the same in upstroke and downstroke (symmetrical). Bats change the areas of their wings between upstroke and downstroke so the wings are not symmetrical between the two motions. On the downstroke bats fully extend their wings and generate most of the lift. On upstroke, the bats partly retract or fold their wings, generating almost no lift. In this phase they adjust the angle of attack to produce enough force to keep it in position.

Hovering is expensive but provides a stable platform for feeding. Nectar is a high-energy food. Specialized digestive physiology allows a Pallas's Long-tongued Bat to fuel about 78% of the cost of hovering with just-ingested sugars. This makes them more like hummingbirds (95%) than humans, who can fuel about 30% of exercise with sugar eaten just before exercise. Sugar concentrations in nectar of bat-pollinated plants range from 3% to 33%. Experiments with captive Lesser Long-nosed Bats revealed that they can distinguish among solutions that differ by 0.5 % sugar concentrations, allowing bats to identify the highest sugar concentrations available.

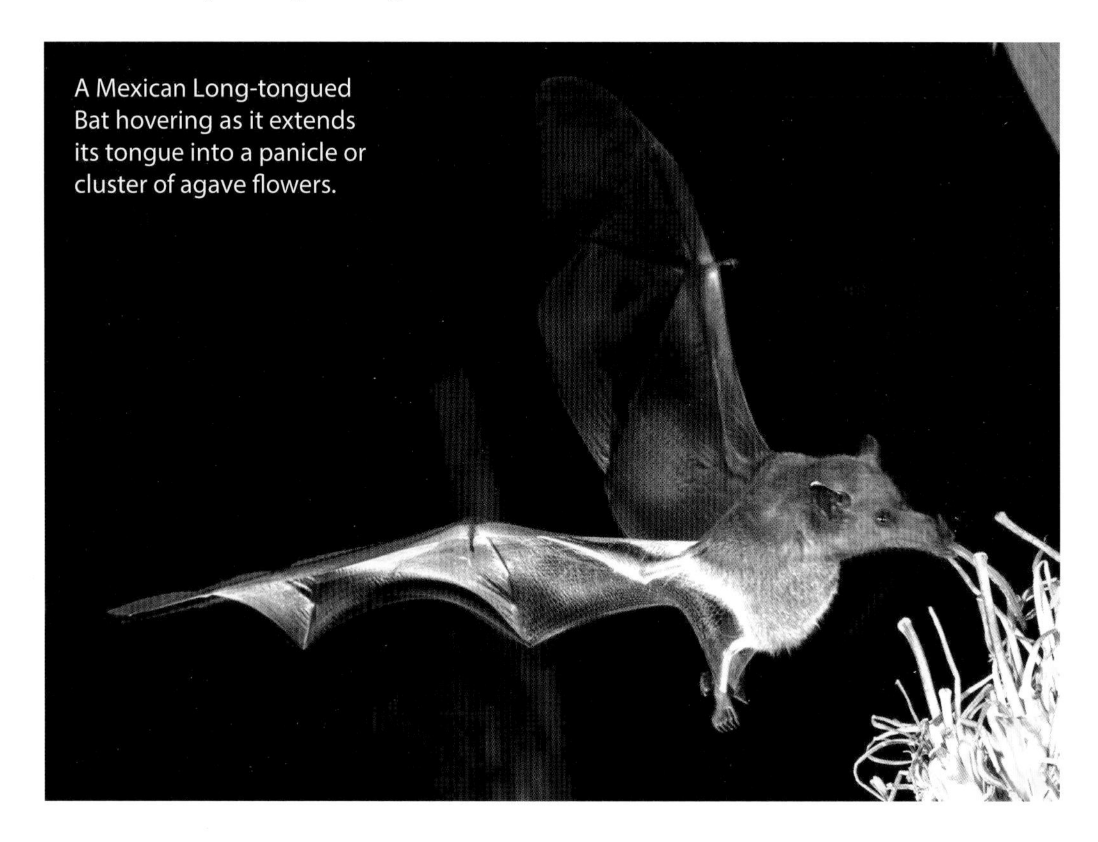

A Mexican Long-tongued Bat hovering as it extends its tongue into a panicle or cluster of agave flowers.

Two views of a Lesser Long-nosed Bat illustrate tongue flexibility. The bat is drinking sugar water from a glass tube.

Collecting nectar is another challenge. The tongues of long-tongued bats are highly flexible (see pages 99–101).[58] Papillae on the tip of the tongue become erect when engorged with blood and soak up nectar. Some other nectar-feeding bats move nectar through grooves in the sides of their tongues. There are at least three evolutionary lineages of nectar-feeding bats in the Americas.[59] In the Old World, some specialized pteropodids are nectar-feeders.

Finding nectar is another challenge for pollinators. But chiropterophilic flowers are conspicuous, reflecting benefits associated with pollination. In addition to nocturnal opening, light colour and strong musty odour, some bat-pollinated plants have beacons – specialized leaves that generate strong echoes of bats' echolocation calls. Even the bracts (specialized leaves) of banana flowers are acoustic beacons, indicating where the bats can find nectar.[60] Banana flowers typically are consistently open for several nights, but at any flower, the location of nectaries varies from night to night. By removing bracts at different times, researchers showed that these structures made it easier for bats to find the nectar on any given night. Bats visiting flowers are usually carrying pollen when they leave.

Mexican Long-tongued Bats and Lesser Long-nosed Bats are not territorial, at least during migration. Several individuals can simultaneously arrive a one feeder or agave flowers, sequentially drinking from either source. The bats clearly travel in groups. Although successful reproduction of many flowering plants depends on pollinators, the animals are less committed. Pallas's Long-tongued Bats, for instance, also eat fruit and insects. Pollen is rich in protein but difficult to digest. Some nectar-feeding bats are said to drink their own urine to foster an acidic environment in their stomachs. This would facilitate digestion of pollen and release of proteins.

Pallas's Long-tongued Bat, its head covered in pollen, entering a night roost in a cistern in Belize.

A cluster of Lesser Long-nosed Bat pups roosting in a lava cave roost in the Sonoran Desert. The sea of pink bat pups engulfs a few mother bats. *JR*

An adult Lesser Long-nosed Bat feed at the flowers of giant Saguaro cacti, which provide nectar and pollen. *JR*

The curious case of bananas

Bananas originated in Southeast Asia. They have been a staple for humans for at least 4,000 years. Over 500 varieties are recognized, but the main crop plant is a sterile hybrid. Banana flowers produce copious amounts of nectar but no pollen. Many animals, including bats, visit banana flowers and consume nectar, but the plant does not appear to benefit from these visits. In the Old World tropics, where there are other species of bananas, specialized pteropodid bats visit flowers and pollinate the banana plants.

Close-up of a Pallas's Long-tongued Bat's tongue with engorged papillae. This makes the tip a nectar mop.

A Pallas's Long-tongued Bat drinking nectar from a banana flower. The arrow points to a bract that serves as a beacon, reflecting echoes of the bat's echolocation calls.

The human market for bananas has generated large plantation operations in many parts of the tropics and subtropics.[61] There are both conventional monocultures and organic banana plantations. Recently, researchers working in Costa Rica showed that Pallas's Long-tongued Bats feeding in organic banana plantations and monocultures were larger (in terms of their body mass) than those in natural forests. This difference probably reflects food density and availability. But bats foraging in conventional monocultures had less diverse and perhaps ineffective gut microbiota. Those foraging in organic plantations had gut microbiota more similar to those of bats in natural forests. This gut microbiota situation is reminiscent of what we see in humans that have different sorts of diets too.

7 Vampire bats

A Common Vampire Bat emerging from its cave roost in Belize

These bats take the name 'vampire' from human mythology where a vampire is a person who comes back from the dead to feed on the blood of the living. In this tradition, European explorers arriving in Central and South America called the blood-feeding bats they found there 'vampires'. Common Vampire Bats are the best known of the three species. They occur in Mexico and south through Central America and much of South America. Fossil vampire bats are known from northern California and Florida, as well as from Cuba and Venezuela. There are no records of blood-feeding bats from anywhere else in the world (including Transylvania).

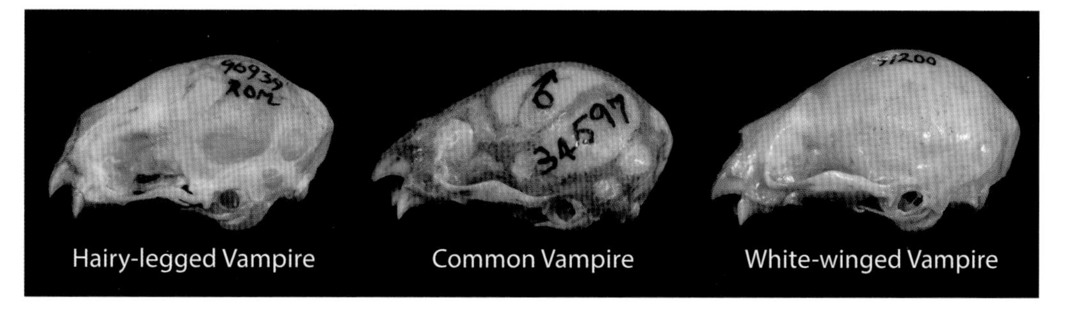

Skulls of the three species of vampire bats (the numbers and symbols identify them in the collections of the Royal Ontario Museum). Each skull is about 2.5 cm long. Among these species, the Hairy-legged Vampire is the most distinct. All three species make feeding bites using their enlarged upper incisor teeth. Feeding bites involve removal of a divot of skin about 5 mm deep and 5 mm in diameter.
Opposite: A Common Vampire Bat partly folds its wings as it flies out of a hollow tree roost.

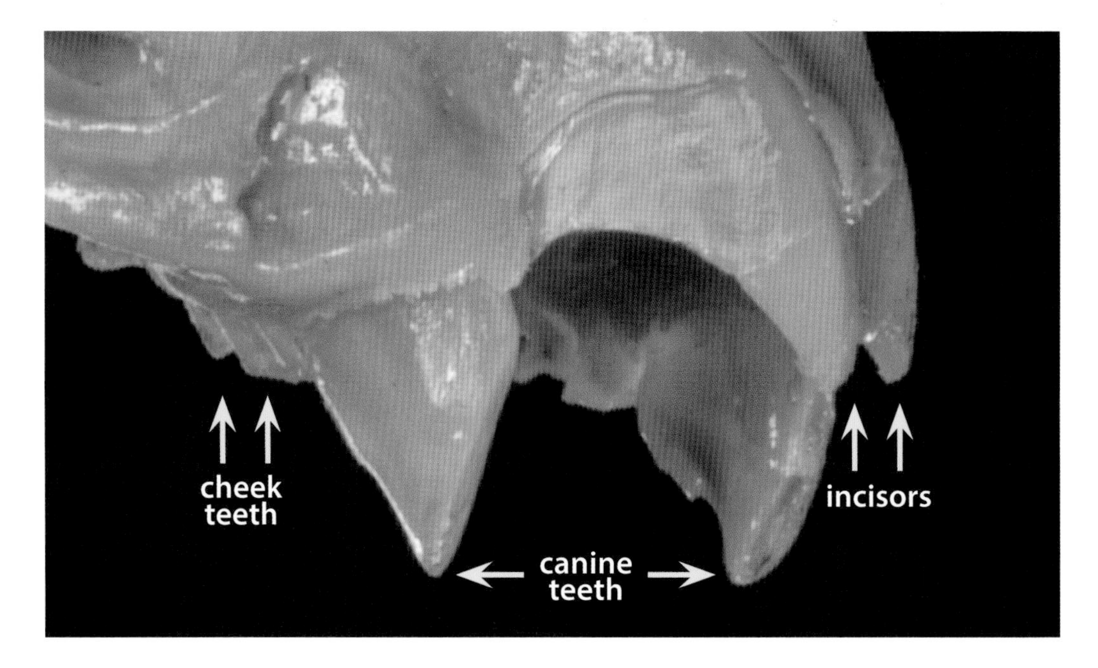

Common Vampire Bat upper teeth and lower teeth. Above, the two upper incisors are shown, along with the two canine teeth, and two cheek teeth. Below, the two lower incisors are shown, a lower canine, and cheek teeth.

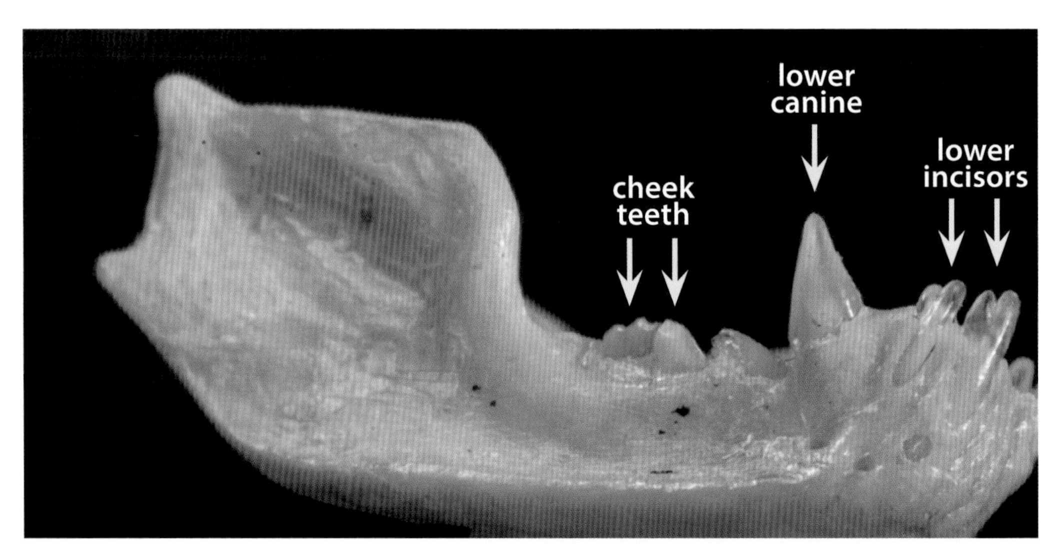

In addition to echolocation, the hearing of Common Vampire Bats is also particularly sensitive to the deep breathing sounds of a sleeping prey. These bats have heat sensors on their noseleafs which permit them to locate places where blood flows close to the skin surface. The bats use specialized cheek teeth in upper and lower jaws to clip away fur and feathers and expose a bite site. The bats do not sever blood vessels, but rather drink blood that flows from the scrape-like bite. Streaks of blood identify vampire bite sites.

Common Vampire Bat

These bats' saliva contains an anticoagulant, 'Draculin'. This inhibits blood clotting and facilitates feeding. Draculin has medical potential because it can break up blood clots such as those resulting from a stroke. The saliva of Hairy-legged Vampire Bats contains mutated forms of a gene that regulates blood pressure.

Vampire bats drink about 25 grams (two tablespoonsful) of blood at a feeding. To reduce the mass of the blood meal, they begin to urinate within two minutes of feeding. The urine is dilute, basically the plasma from that blood meal. Even so, vampire bats may have difficulty taking off with a full load of blood. They launch themselves into the air by contracting the muscles that power the downstroke in flight.

Traces of blood (indicated by an arrow) reveal that during the night at least one Common Vampire Bat fed on the neck of this cow.

On the ground, Common Vampire Bats are quick, agile and nimble. The bats' running gait involves first anchoring the hind feet and extending their wrists forward, then anchoring the wrists and thumbs and swinging their bodies and hind feet forward. The long forearms and thumbs translate into a long stride compared to four-footed mammals of the same size (such as mice). Running Common Vampire Bats raise their metabolic rates 1.5 times. This is a much lower cost of running than seen in mice (4.8 times). During running their body temperatures increase from about 38°C (100.4°F) to just over 42°C (107.6°F). In other words, the gaits of running Common Vampire Bats are far less expensive in metabolic terms than running in mice. Clearly, they are well adapted to moving around in this way, but as yet we do not know the full significance of this mobility and agility on the ground.

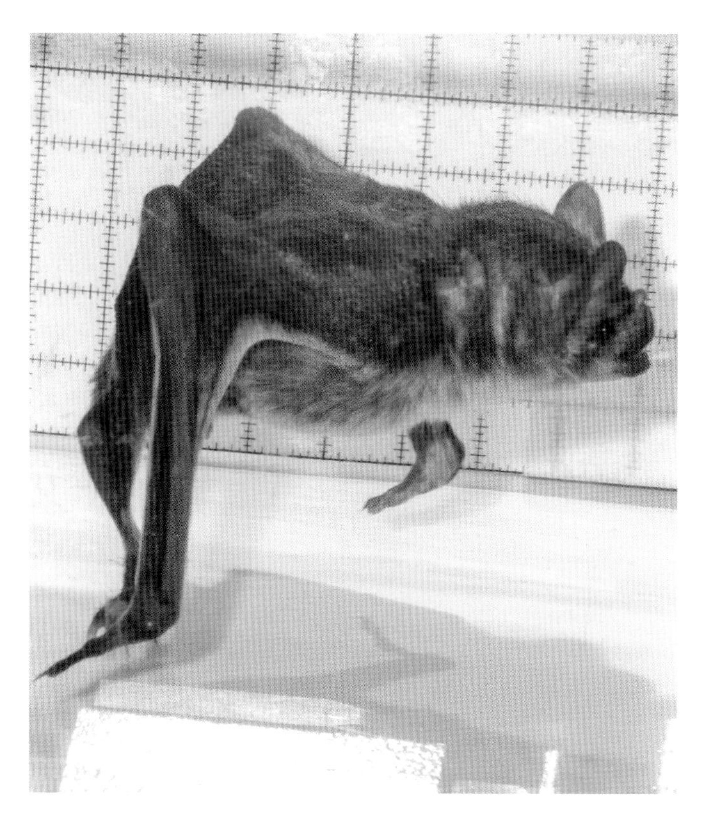

Two views of a Common Vampire Bat running on a treadmill. On the top, the bat has one hind foot down just behind its right wing and is reaching forward with its wrists. On the bottom the bat has planted its wrists and is bringing its hind legs forward.

Vampire bats can only last one night without a blood meal. Adult vampires typically fail to feed on average of one night a month, while young animals, presumably because they are less experienced, may fail to feed two or three nights in a week.[62] The social network in colonies of vampire bats is reflected by successful bats regurgitating blood to their colony members who failed to feed. But dependence on blood could mean higher risks of starvation. In areas with abundant livestock, populations of Common Vampire Bats may reproduce more often. This could reflect higher rates of survival as well as more opportunities to reproduce, where variation in rates of survival could reflect local conditions.

Mayan societies occupied much of Central America (southern Mexico, Guatemala, Belize and western Honduras and El Salvador) for over 3,000 years. Mayan people did not depend upon grazing mammals as food, rather on plant crops, fowl, smaller-bodied mammals and various marine or freshwater resources.[63] Europeans brought larger grazing mammals which, with expanding populations, would have changed feeding opportunities for vampire bats. In Mayan times, humans would have been the largest mammal food sources for the bats – and indeed, in some parts of Brazil and Peru, Common Vampire Bats routinely consume human blood. Some people in Belize place branches of the Tiger Nail Plant *Martynia annua* over the open parts of chicken coops: the strong thorns entangle the bats and protect the chickens from their raids. Other people keep bats out by hanging bunches of marigold flowers or fruits of wild pine, but the basis of these plants as bat repellents is not clear.

In the tropics of Central and South America, Common Vampire Bats are an important reservoir for the strain of rabies that afflicts these bats.[64] Research in Belize (Central America) revealed widespread evidence of exposure to rabies in this species. This suggests endemic circulation of the virus. In our study area in Belize, we found occasional movements of Common Vampires among roosts 6 km apart. This could facilitate the continued spread of rabies among the wider population.

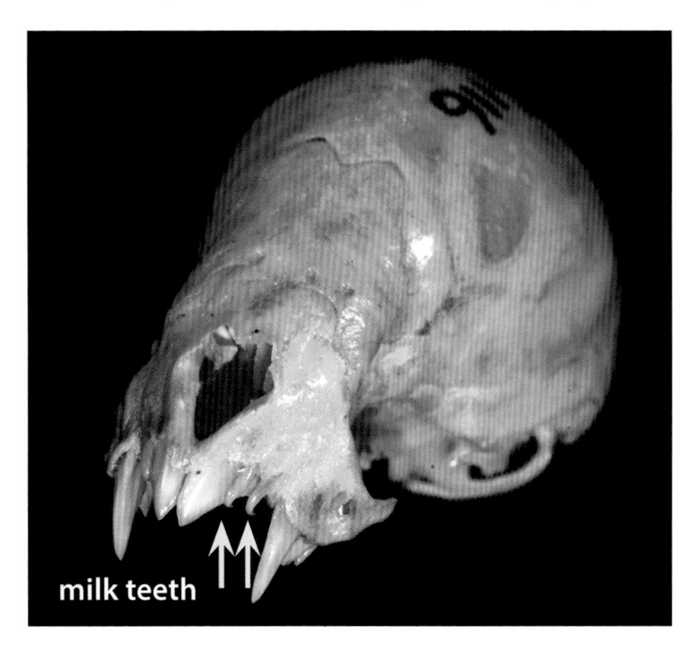

Skull of a Common Vampire Bat showing deciduous (milk) teeth. Note the differences between the slender, slightly hooked milk teeth and the broader, flatter permanent teeth. Hooked milk teeth could facilitate latching onto the mother and feeding.

Mother Common Vampire Bat nursing a pup, whose wrist and ear can be seen.

Like other bats, Common Vampires are long-lived (over 19 years in the wild) but reproduce more slowly than many other species. Dependent young nurse for about nine months, limiting annual reproductive output. Tooth sharpness in vampire bats will prove very interesting. How do the bats keep their teeth (especially upper incisors) sharp? And how do female vampires avoid being sliced up when nursing young that already have adult dentition?

In 2022, Moritz Blumer and his colleagues examined gene losses (indicating evolutionary adaptations) associated with blood-feeding in vampire bats.[65] Compared to 26 other species of bats, ten gene losses have occurred in Common Vampire Bats and these may underlie specializations for blood-feeding. The genes involved affect insulin secretion, glycogen synthesis and gastric physiology. Collectively, they reflect differences in digestive functions compared to other bats. Urinating just after starting to feed could be an outward indication of the genetic changes.

A Canadian Dracula stamp from 1997 recognizes the supernatural associations of bats.

When it comes to folklore, vampire bats have captured people's imagination. The connection of vampire bats to Dracula appears to reflect the fact that these animals were in the news when Bram Stoker was writing his famous novel in the 1890s. In human mythology, a vampire is a person who returns from the dead to drink the blood of living people. The notion of bats being associated with vampires is reflected by the names of several species (see page x). But a young, unweaned Common Vampire Bat is hardly a fearsome sight.

A juvenile Common Vampire Bat epitomizes the cuteness of many young animals.

As blood-feeders, vampire bats can be vectors for rabies (see page 174) and have an economic impact on livestock farming in some parts of South and Central America. Common Vampire Bats are more often a problem than either Hairy-legged or White-winged Vampire bats, which could reflect their relative abundances. Vaccinating cattle against rabies can be an effective way to solve the problem of vampire bats spreading the disease. Wholesale killing of vampire bats in their roosts is another approach, by pasting captured bats with an anticoagulant such as warfarin. This very effectively kills individuals that ingest the paste because vampire bats that roost together often groom one another, thus spreading the paste and killing more bats. Sometimes people that catch the bats also break the animals' teeth keep them from biting.

Control measures that require catching bats are ineffective when the people involved do not know how to recognise a vampire bat. To put the problem in perspective, do you think that any of the bats below are vampires?

Are any of the six bat species pictured here vampires? If so, which ones? How did you recognise them? (See page 235 for the answer.)

8 Where bats occur and where they roost

Although bats are mainly animals of the tropics and subtropics, they occur on all continents except Antarctica. Bats occur on many remote oceanic islands such as Fiji and Christmas Island. Bats of course require food and roosts, secure places to spend the day. These animals are bite-sized for many predators, but their small size can make it easy for them to hide. We do not know where many species roost. This could mean we do not know the roosts of 700 of the roughly 1,400 species. How do you know which species use a roost? We use a variation on a camera trap,[66] which minimizes disturbance and can be effective when we cannot enter a potential roost.

One of the entrances to St Clair Cave in Jamaica, photograph taken from below and showing dangling tree roots. At least eight species of bats roost in this cave, making it one of the most important known sites in the country.

African Sheath-tailed Bats flying about and roosting in a cave in Kenya (opposite). Other caves or equivalent features (see, for example, page 121) harbour thousands of bats but none is visible at the entrance during the day.

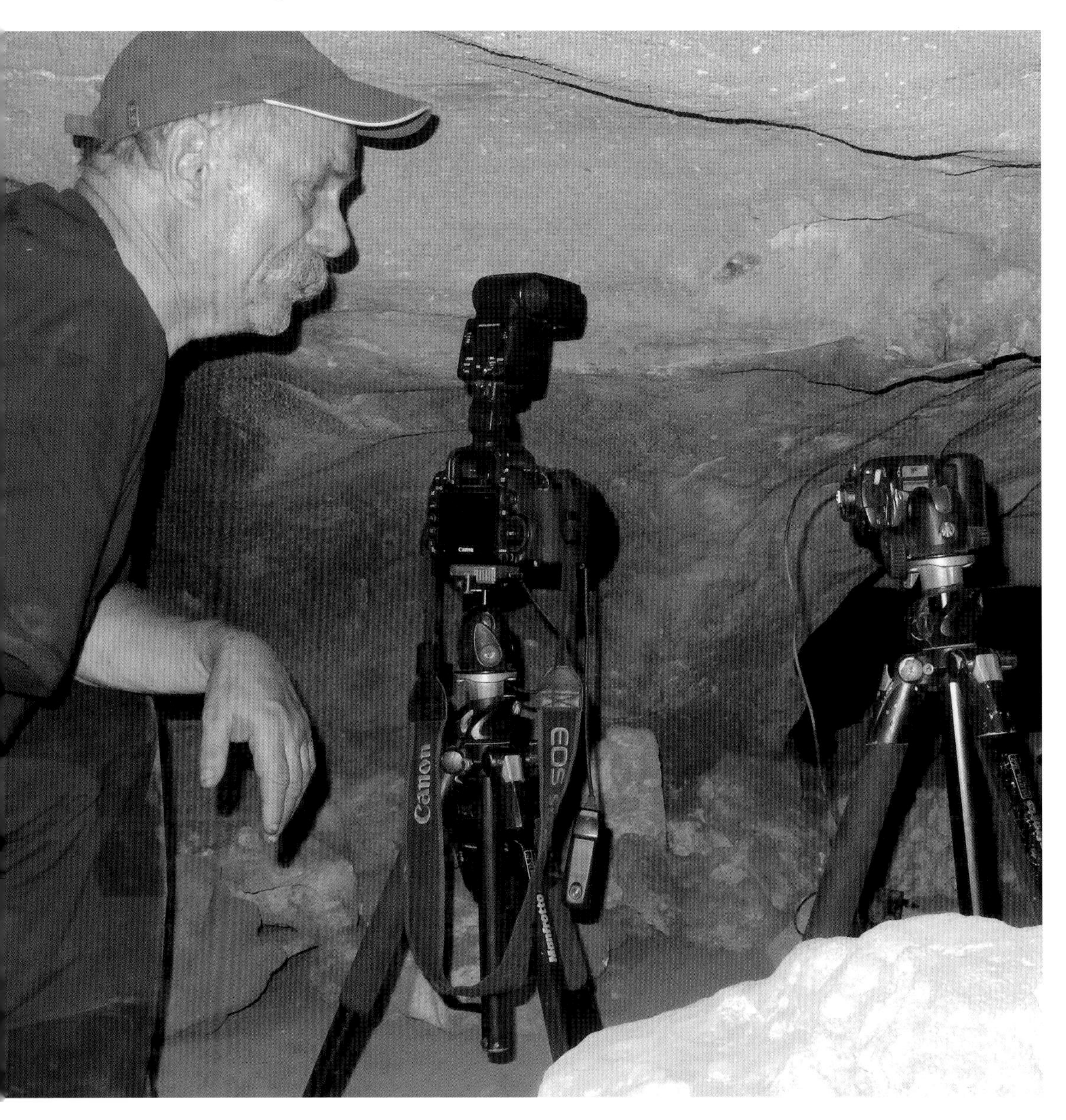

Jens Rydell at his photo setup in a cave in Cuba. Tripods hold a camera and flashes. Flying bats break the infrared beam and trigger the system. Short duration flash(es) freeze bats in flight.
Photo by Brian Keeley

Bats often roost in artificial sites such as abandoned mines and military bunkers, not to mention buildings and bridges (see pages 139 and 171, for instance). In southeastern Ontario (Canada) there are two such abandoned mines in the Canadian Shield. Each was opened in 1906, closed in 1918. Before the arrival of white-nose syndrome (see page 191), these were hibernation sites for over 80,000 Little Brown Myotis. Both sites have been used by bats for over 100 years. There are many other important sites for bats that are the direct result of human activities.

MBF standing outside an abandoned mine, a hibernaculum for bats such as Little Brown Myotis. Outside it was -30°C (22°F), but inside behind the ice curtain there was a constant temperature of about 4°C (39.2°F). At this time (February 2020) we counted over 4,500 bats hibernating at the site. Before the arrival of white-nose syndrome in 2010 there had been over 50,000 hibernating there (see page 172). In summer, Little Brown Myotis roost in warm places such as the attics of buildings.

Temperature

Body size influences bats' temperature requirements, and bats often use different roosts in different seasons. These animals use a great deal of energy, so roosts suitable in winter may not work in summer. Roosts can provide favourable thermal environments for bats, shielding them from extremes of temperature. Many temperate species of bats have variable thermostats,[67] allowing their body temperatures to fall to ambient when they return from foraging. Indeed, some bats can tolerate their internal body temperatures varying by tens of degrees. They can enter torpor, a period of reduced metabolism which can mean temperatures just above freezing. Hibernation consists of extended periods

Hibernating Little Brown Myotis covered in condensation

Northern Bats often hibernate in cold rooms in basements. *JR*

of torpor, ranging from days to even weeks. A Little Brown Myotis hibernating at 4°C (39.2°F) has a heart rate of 5 beats per minute. However, the bat can burn fat and, in 60 min, raise its body temperature to over 35°C (95°F). This consumes as much energy as it would burn in 60 days of hibernation. Other hibernating mammals, such as ground squirrels, usually arouse from torpor once every week or so. Hibernating bats can go up to 60 days without waking up. Stored food is one difference between rodents and bats: hibernating ground squirrels cache food (plant material) in their burrows and, after stirring from sleep, eat enough calories to cover the costs of being awake. Hibernating bats do not have food caches and must rely entirely on their stored body fat. Some bats hibernate in buildings; this may be above insulation or below ground, spaces where temperatures are just above freezing. In the United States and Canada, Big Brown Bats commonly hibernate in buildings but, for the most part, the exact hibernation sites are not known. Hibernating bats are sometimes found during renovations when the walls are opened. The bats are always protected from cold (outside) and heat (inside). People frequently encounter such bats when they come out of hibernation and search for water.

Larger species face different challenges associated with roost temperatures. Egyptian Fruit Bats (see pages 5–6) are about ten times heavier than Big Brown Bats. The situation in a cave on Table Mountain (South Africa) reveal the temperatures Egyptian Fruit Bats must sometimes endure. Researchers used radio tags that measured bats' skin temperatures.[68] Within the cave, the ambient temperature ranges from 7°C (44.6°F) to 12°C (53.6°F). The body temperatures of bats returning from foraging average 37.7°C (99.9°F), dropping to 35.5°C (95.9°F) by noon. Egyptian Fruit Bats remain active in the cave during the day. Their larger body sizes means that it costs them less to stay active than it would have cost Big Brown Bats.

In summer, Big Brown Bats (14–20 g) roosting in an attic faced a different temperature regime.[69] The same types of tags that measure skin temperature revealed that, on returning from foraging, bats' body temperatures averaged 33°C (91.4°F), when roost temperatures ranged from 8°C (46.4°F) to 28°C (82.4°F). During the day some individuals maintained their body temperatures from 30°C (86°F) to 35°C (95°F) until they emerged to forage again. The body temperatures of others dropped to about 12°C (53.6°F) within 30 minutes of their return and then followed the roost temperatures through the day, increasing to around 30°C (86°F) just before emergence. Big Brown Bats typically stop producing metabolic body heat when the air temperature drops below about 26°C (78.8°F), and their body temperatures can fall to 1.1°C (34°F).

Big Brown Bats, like many other species, use periods of torpor to reduce the energetic costs of temperature regulation. This allows them to exploit otherwise unavailable areas in which to roost. Pregnant females make less use of torpor than adult males. Pregnant female Hoary Bats (30 g) used torpor during spring storms, emerging from sleep and giving birth when the weather improved.[70] Reduced body temperatures during pregnancy prolong the gestation period. This approach ensures that lactating females face improved conditions for their dependent young.

A wide body of evidence indicates that smaller bats cool down faster than larger species.[71] In Belize, for example, most of 99 non-reproductive bats of 13 species (four families) allowed body temperature to drop under cooler conditions. The sample size in this work did not make it possible to determine (statistically) the relative importance of diet (fruit versus insects) for thermoregulation.

The move from summer activity to reducing levels of activity through late summer and early autumn reflect changes in weather as well as availability of insect prey. Researchers studied two groups of captive male Particoloured Bats, one on a restricted food diet, the other not so restricted. The results showed that consistent body masses occurred when food intake was restricted. Males on unrestricted diets gained body mass, while there was little difference in use of torpor between fed and unfed bats. But, compared to bats in the unrestricted group, sexual maturity was delayed by half a month in food-restricted individuals.

Nursery colonies of bats, females and dependent young have more stringent requirements of roosts. Usually these bats require places with shelter, enough warmth to minimize costs of thermoregulation, and protection from predators. For hibernating bats – male or female – roost temperatures must not fall below freezing, and for many species high humidity is crucial. The Fishing Myotis (25 g) of Baja California roost in rock cavities and crevices that provide shelter and better insulation in summer than in winter.[72] Where these bats occur, winter temperature can be 5°C (41°F), then rise to as much as 45°C (113°F) in summer. In these roosts, radio transmitters that measured skin temperature revealed that individual bats made variable use of torpor. In 2020 researchers from Mexico measured the responses of Fishing Myotis to different levels of environmental stress. In late winter researchers found lower levels of a hormone associated with stress in the bats' faeces. They encountered higher levels of stress hormones in summer and early winter. So, in challenging roost environments these bats maintain constant basal glucocorticoids but respond differently to acute stress.

One of the most interesting discoveries about hibernating bats came from areas of heavy snowfall in Japan in 2018.[73] Researchers there found Ussuri Tube-nosed Bats (4–5 g) hibernating under snow. When discovered the bats were lethargic, presumably torpid. Use of snow for hibernation sites appears widespread in Japan. The researchers suggested that these bats might initially roost in foliage, but with the arrival of colder weather move into hollows in trees. In much colder weather, snow may offer better insulation for a hibernating bat than a tree hollow. In autumn in some parts of the American Midwest, radio-tagged Eastern Red Bats have been found hibernating under the leaf litter.

For many individuals, roost structure itself may be less important than the other bats occupying it. Under cold or cool conditions, roosting bats often cluster together to benefit from their collective body heat and minimize heat loss across exposed surface areas. Roosting with conspecifics, Big Brown Bats reduce their costs of temperature regulation by about 60%. In summer, Lesser Short-tailed Bats in New Zealand used communal thermally stable roosts. In winter, roosts of solitary individuals were warmer than the ambient temperature. Some bats clearly change roosts depending upon the situation.

In Sweden and elsewhere in northern Europe, Soprano Pipistrelles sometimes hibernate above insulation in attics. *JR*

Bat roosts

Some bats roost in hollows or crevices, which may be in rock (i.e. caves) or trees or buildings. Others roost in foliage or, less commonly, under rocks on the ground. A few species roost in plain sight, where they can be easy to overlook. Searching for roosting bats is usually not productive, although placing a radio tag on a bat can allow you to follow the animal to its roost.

Above: A colony of Straw-coloured Fruit Bats (250 g) roosting in plain sight on a large tree. *JR*
Below: A Greater Long-fingered Bat coming out of a cave in Africa. *JR*

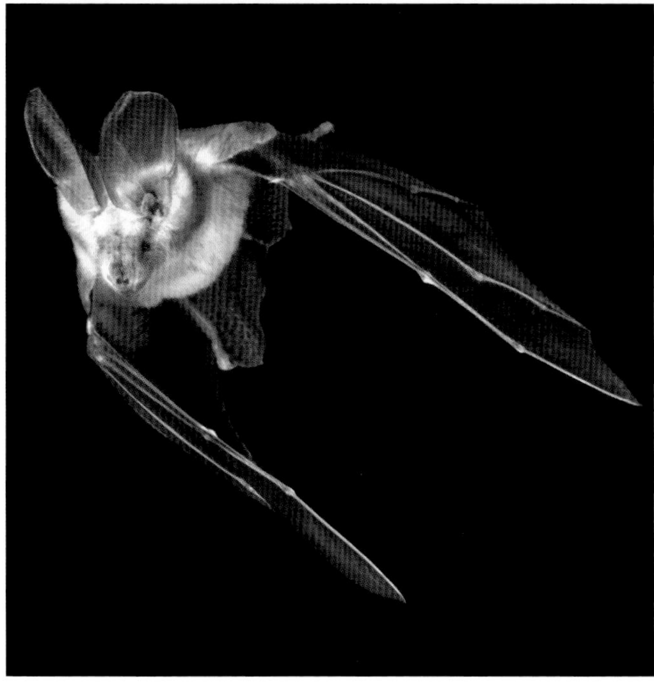

Above: a Heart-nosed Bat emerging from a well. *JR*
Left: an Egyptian Slit-faced Bat exiting a cave roost.

Proboscis Bats roosting by day on the trunk of a Guanacaste tree in Belize. Like other sheath-tailed bats this species has a characteristic roosting posture, hind feet and both wrists on the substrate. The closer view above reveals characteristic tufts of fur on the bats' forearms. Note the spaces between individuals.

Here are examples of some different bat roosts. Straw-coloured Fruit Bats hang in the open, sometimes on trunks or main tree branches, other times in foliage. Other bat species roost in wells. In Gede on the Kenyan coast, some boys told JR about bats roosting in a nearby well. The camera system documented about eight Heart-nosed Bats (30 g) emerging. Larger roosts such as caves often harbour several species – for instance Greater Long-fingered Bats (15 g) or Egyptian Slit-faced Bats (10 g). Photographic surveys are low-impact ways to accurately identify which bats are using a roost.

Proboscis Bats are small (4 g) and well camouflaged; this species also roosts in the open in plain sight. Observations with a thermal-imaging camera revealed that they moved around the trunk of a palm tree, staying in the same thermal zone (i.e. keeping in the relatively cool shade as the sun moved across the sky). Temperature conditions usually explain degrees of body contact among roosting bats. Individual Straw-coloured Fruit Bats may roost in close contact with others (clusters), with spaces between clusters.

Sundevall's Roundleaf Bats roosting in Kenya. Sometimes roosting bats are spread out or form tight clusters (see over). *JR*

When you have a light, bats roosting in caves may be in plain sight. In caves and hollows, bats roosting on the ceiling can be spread out, such as Sundevall's Roundleaf Bats who prefer to be evenly distributed on the roost surface. Other species such as Natal Long-fingered Bats (10 g) sometimes form tight clusters, to share body heat. This is particularly important for very young individuals. During hibernation, clustering may also minimize water loss.

Wahlberg's Epauletted Fruit Bats roosting in the foliage of a tree in Kruger National Park (South Africa). The bats are inconspicuous because of the combination of keeping quiet and still as well as disruptive colouration achieved by white patches of fur at the bases of their ears. How many bats can you see in the roost?

Opposite: Natal Long-fingered Bat roost in South Africa. For more on this species see page 140. The pink-coloured features on one of the bats towards the bottom left are streblid flies which have buried into this individual's face – these parasitic insects usually live in a bat's tear ducts (see page 177). *JR*

Patterning in bats

Some tent-making phyllostomid bats (see pages 136 and 137) have white or light-coloured stripes on their faces. Here, Heller's Broad-nosed Bat has four facial stripes and one stripe up the middle of the back. A Pygmy Fruit-eating Bat and a Thomas's Fruit-eating Bat each have four facial stripes but no dorsal stripe. Males do not appear to differ from females in these features.

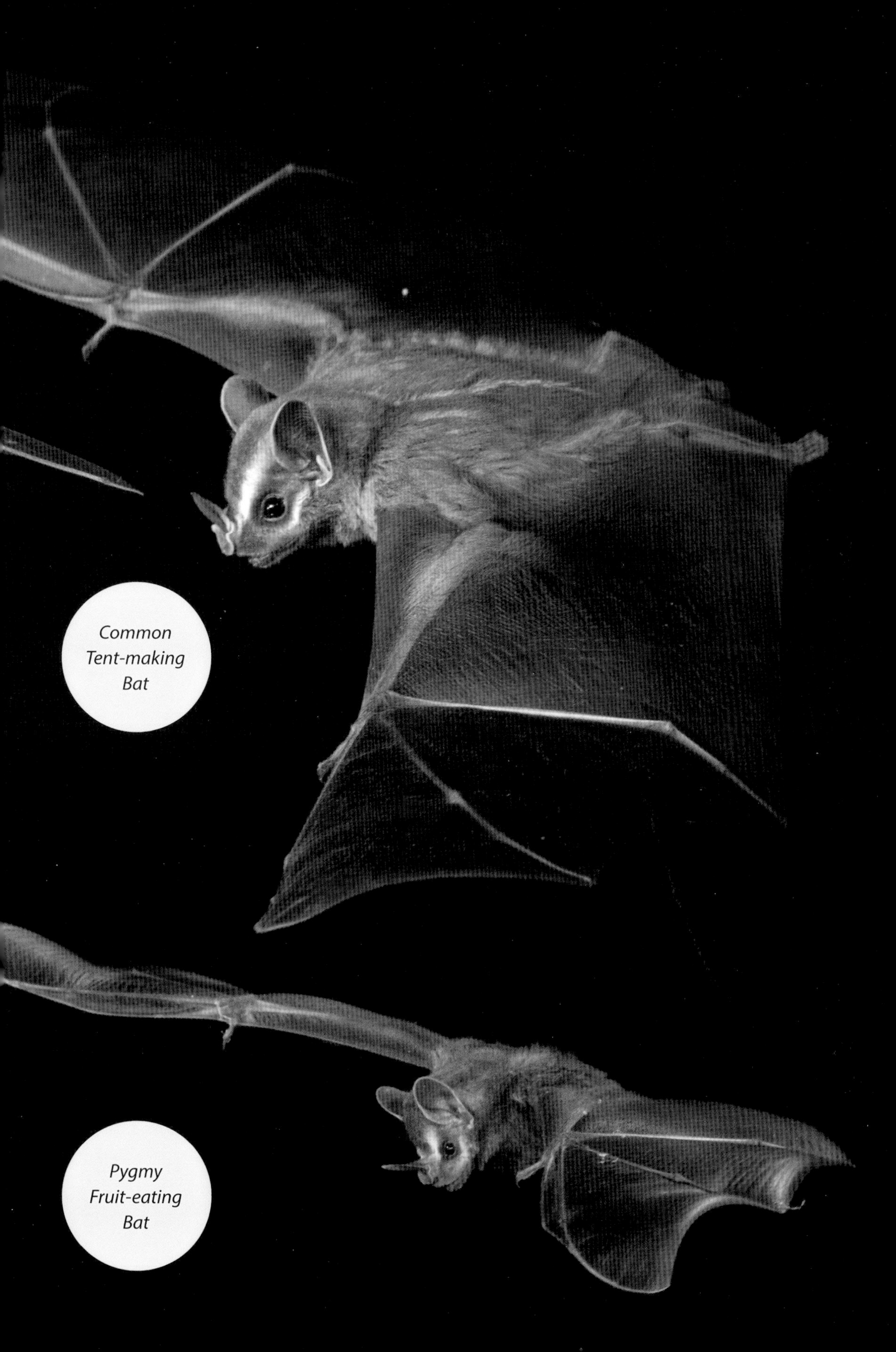
Common
Tent-making
Bat
Pygmy
Fruit-eating
Bat

Bats make different uses of foliage as roosting sites. Some species roost among foliage, others in furled leaves. Disks on their wrists and ankles facilitate movement on slippery leaf surfaces. In the New World tropics, several species of disk-winged bats use suction to adhere to leaf surfaces. In Madagascar two other species from a different family use adhesion.

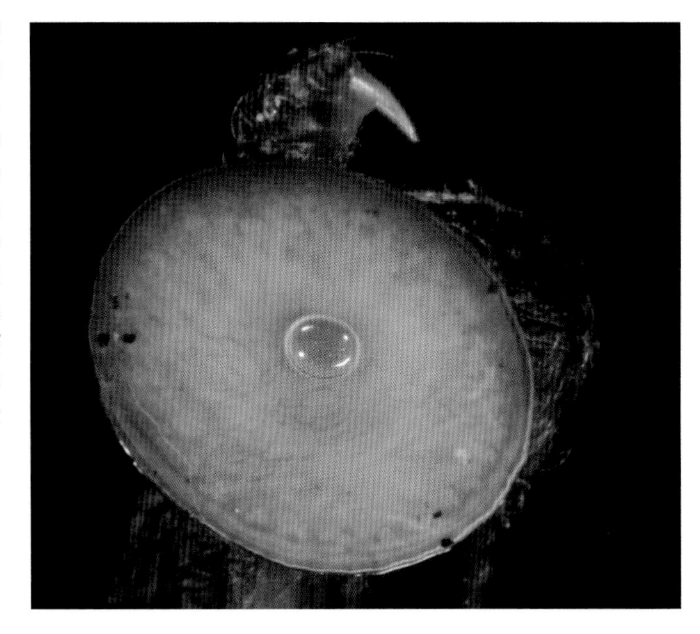

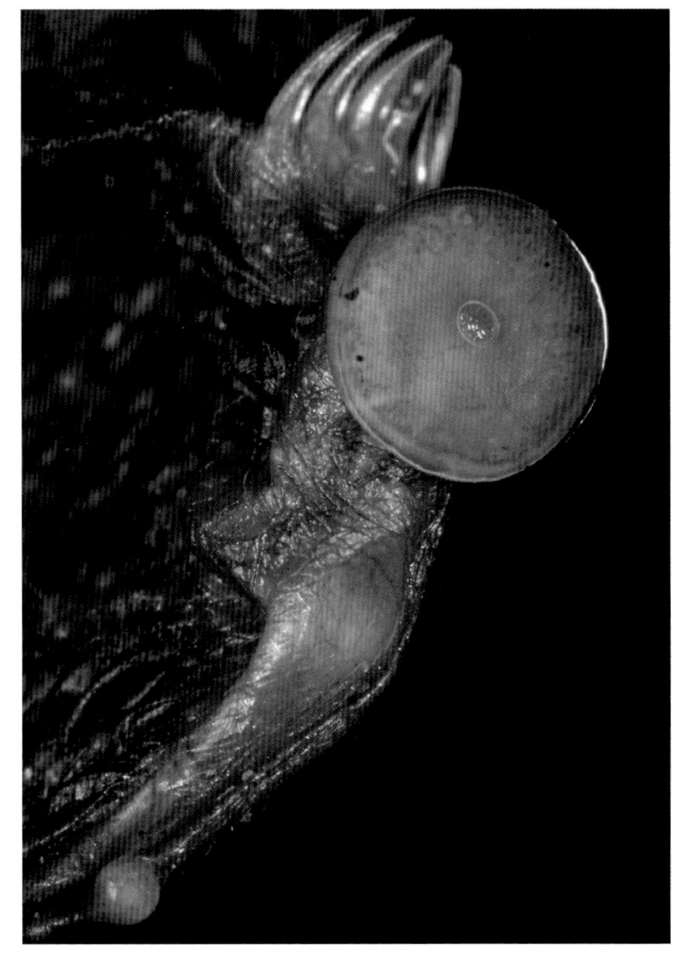

In the upper image a disk is shown with the thumb. In the other view, the bat's ankle disk, hind feet and calcar (a spur of cartilage that reinforces the wing) also are visible.

The silhouette of a Spix's Disk-winged Bat roosting in a furled leaf. The bat's wrist disks are clear in the silhouette.

Side view of a Heliconia leaf modified as a bat tent (above). Such structures are used by various species, such as a pair of Pygmy Fruit-eating Bats (below), two Yellow-eared Bats under a philodendron leaf (opposite top) or a cluster of Honduran White Bats (opposite bottom).

Although bats do not build nests, some species do make tents. Tent-making remains one of the most remarkable aspects of bat behaviour yet to be discovered.[74] Although best known from the New World tropics, and from phyllostomid bats specifically, the same sort of behaviour occurs in some smaller species of pteropodids in India and Malaysia.

Tent-making species bite leaves, causing them to fold over. The resulting space under the modified leaf protects the occupants from direct sunlight and rain. The bats' bites damage vessels in that supply water and nutrients to the leaf, but despite this some tents can last from 45 days to a year. Researchers in Costa Rica used dyes to follow water movement in leaves modified as tents. What botanists had considered minor veins were not damaged and continued to serve the plant. Maintaining water and nutrients meant a green tent, which is much less conspicuous than a dead section of a leaf. Tents can be important to bats, and systematic sampling revealed that a succession of species uses some tents. It appears that in certain cases tents are constructed by males, but for most bats there are few data about tent construction.

A nursery colony of Brown Long-eared Bats in an attic roost. The bats exploit the warmth of the roost and their roost mates to maximize growth rates of their young. Female bats are very choosy about nursery roosts and quickly abandon them after disturbance. The accumulation of discarded prey parts (see also page 67) provides yet more information about the bats. *JR*

Many, perhaps most, species of bats roost in groups, usually all members of the same species. Groups ('camps') of flying foxes are conspicuous because of the noise and wing flapping that characterize these aggregations. Smaller pteropodid bats (~100 g), such as Wahlberg's Epauletted Fruit Bats, roost in trees (see page 131). Radio-tracking has allowed us to locate the specific tree or bush in which a tagged bat is roosting. This worked with Wahlberg's Epauletted Fruit Bats (~100 g), Eastern Red Bats (~10 g), Hoary Bats (~30 g), Wrinkle-faced Bats (~20 g) and Northern Yellow-shouldered Bats (~18 g). But, in spite of strenuous effort we could not spot roosting bats which were still, quiet and so virtually invisible. They are, however, alert and quick to flush if disturbed. For example, a radio-tagged Northern Yellow-shouldered Bat was roosting in a shrub about 5 m tall. We could not see it, but when we tapped a branch, six of these bats took off and flew away.

Warm to hot roosts with little air flow that are used by vast numbers of bats pose special challenges. One is accumulation of ammonia associated with bat urine and faeces. Another is a shortage of oxygen. Witness a roost in an abandoned kaolin mine in Tanzania, where there was not enough oxygen to support the flame of a pressure lamp. The combination of ammonia and a shortage of oxygen makes some roosts inaccessible to bat biologists. As we shall see later (page 211), bat roosts can be places where people may be exposed to histoplasmosis, a fungal disease of the lungs.

We know that bats adjust their choice of roost depending upon conditions. For species in temperate areas, temperature profiles in winter roosts differ from those in summer roosts. Night roosts are places where bats rest and/or eat between foraging bouts. They may not be the same as day roosts. Radio-tracking revealed that some bats use a variety of roosts, and regularly (or unpredictably) switch roosts. Bats may change roosts in response to disturbance or predation events; they may also may have to deal with loss of roosts.[75] In Cypress Hills Interprovincial Park in Saskatchewan, Big Brown Bats form nursery colonies in hollows in Trembling Aspen *Populus tremuloides* trees. Over three years, loss of roosts meant that bats had to move 7 km to another area. There are two important messages here. First, bats must 'know' about other roosts. Second, conservation efforts must not only protect bats directly but also their roosts.

During the summer in Arizona, groups of Pallid Bats roost in rock hollows and crevices by day, often moving to a new roost when they finish foraging just before dawn. In these situations, bats locate the 'roost of the day' by listening to the directive calls of roost mates. Spix's Disk-Winged bats roost in furled leaves. These bats must change roosts almost every day, or when the leaf is too open to be useful any longer. These bats also use calls to alert their roost mates to the unfurled leaf of choice on a particular day.

Although tens of thousands of bats might have been banded between 1950 and 1970, the incidence of band recoveries from anywhere other than the site of banding was extremely low. Such small rates of band recovery meant that we learned very little about the movement patterns of bats from this method. Ironically, the arrival and spread of white-nose syndrome in northeastern North America (see page 191) revealed just how much Little Brown Myotis moved among hibernation sites. Using PIT tags (see page 15) to follow the movements of bats gave specific details of how often and when tagged bats moved among roosts.

PIT tags have allowed researchers to document associations among individual bats. Work in Nova Scotia, Canada, revealed associations among tagged Little Brown Myotis and Northern Long-eared Bats. Females in nursery colonies in June and July associated with individuals of the same species. Such within-species patterns also occurred at underground sites during swarming (see page 165). These findings provided further evidence that bats in a colony may change sites and roost-mates. Such 'fission–fusion' social organization (see page 154) also has been suggested for other species of bats. The window that PIT tags provided on bat behaviour has thus greatly expanded our knowledge of the social lives of bats.

In South Africa, PIT tags showed the extent of movements of Natal Long-fingered Bats (10 g) among caves that serve as hibernacula in winter or nursery sites in summer. Individuals with PIT tags move 160 to 200 km in 3–5 days. In this work, some individuals are recorded over a year after tagging. The work also demonstrated the importance of tag size and the reality that not all species are readily studied by PIT tags. Specifically, the skin of Sundevall's Roundleaf Bats and Egyptian Slit-faced Bats is thin, and so some individuals scratched out the tags. The same may apply to smaller species of horseshoe bats.

A Particoloured Bat (15 g), a species that appears to be expanding its range northwards in Scandinavia. *JR*

Lingering challenges

Many observations indicate that the calls of bats attract others. Indeed, people use the calls of bats caught in a net or a harp trap to attract other bats. These records may involve echolocation calls or social calls, or some combination of the two. This means that once some bats find and use a roost, they may be followed by more. But how do the first bats locate suitable roosting sites?

Many bats are opportunistic in their use of roosts. A favourite example is bats roosting in rolled-up parasols. In southern Ontario, this is especially common in the latter part of August and early in September, with the arrival of autumn and at a time when bats disperse. We have also found bats roosting behind shutters on a building and in rolled-up awnings. We have even seen them roosting behind pictures hanging on a wall. But when you go out to find bats roosting, most often you return empty-handed.

Knowing what piques a bat's curiosity might allow us to minimize bat mortality at wind turbines for example. Roosts are vital to bats which often abound in the likes of old buildings and abandoned mines. Bats have expanded their ranges in parts of the world by roosting in artificial underground structures. These include mines, military bunkers and tunnels for moving water (rhettara). In Scandinavia, Particoloured Bats have expanded their range northwards, exploiting roosting opportunities in high-rise buildings.

Bats up north

The number of bat species is highest in the tropics and declines rather abruptly towards the poles. There are more than a hundred species in Kenya at the equator, 44 in the whole of Europe and 19 in Scandinavia.[76] Just south of the Arctic Circle in Yukon Territory in Canada (60°N), Little Brown Myotis are relatively common – as revealed by acoustic monitoring. These bats were as active in low-density urban settings around White Horse and in boreal forest. In Yukon, there are records of four other species of bats compared to nine species in Alberta, south of the Yukon. In comparison, at the Arctic Circle in Lapland there is just one species present, the Northern Bat. Why are there so few bats in the far north? This is puzzling, as the scarcity of bats in these regions contrasts with the abundance of insectivorous birds.

Below: A Northern Bat, flying in Lapland. Opposite: a close-up of the same species. *JR*

Bats in the far north face at least five challenges that do not concern birds, which benefit from the longer foraging time afforded by extra daylight. First, summers in the north are very short. The gestation period for temperate bats is about 60 days, the duration of lactation an additional 25 days. This is about the same length of time as the northern summer, June to August.

Second, bats must put on fat to survive the winter in hibernation, which requires another few weeks of feeding. In the north, winters are long and cold. Hibernation requires a roost that is cool (0–4°C; 32–39.2°F) and humid. But the period of hibernation would have to be long, perhaps eight months or more. The bats hibernate in houses or in rock scree, well insulated by snow.

Third, bats require access to flying insects, usually at temperatures above 10°C (50°F). There are many of those, but only for short periods and when the weather is good.

Fourth, for insectivorous birds, the Arctic days are long, giving them greater access to insects as food. In the Arctic bats have much shorter times to forage. Northern summer nights are very short due to the midnight sun, so bats there must feed rapidly and concentrate activity to the darkest possible sites – in dense forest, on the shady side of the valley or in cloudy weather. Alternatively, they feed high in the open air, probably because they need a good view of approaching predators. Feeding flights take place around midnight, when the sun is lowest, shadows longest and, unfortunately, the air coolest. For a Northern Bat in Lapland, with no protective darkness, the risk of becoming food for a Sparrowhawk is ever present.

Fifth, there is limited access to warm day roosts for bats, which are necessary to speed up the breeding process in summer. In the north, bat maternity colonies have so far occurred only in heated houses. Meanwhile, males have been radio-tracked to cliffs and rock-scree, which is probably where they roost most of the time.

Bat boxes

Around the world, many people have witnessed bats moving into artificial roosts. These usually are in buildings or in underground features such as mines, culverts and military bunkers. Some people also erect bat boxes, some of which are used by bats. Some conservation efforts directed at bats involve putting up bat boxes. In at least one study in Scandinavia, there were very few if any bats in a pine forest on the shore of a lake. The pine trees were too young to have hollows, but after researchers installed bat houses, several species frequented the area and reproduced there.

A 2FN bat box (Schwegler, Germany) with a 50-mm-diameter reflector. Photo by Gerald Kerth

Recent research has revealed that placing an echo-reflector on a bat box attracts individuals of at least three European species, Bechstein's Bat, Natterer's Bat and Brown Long-eared Bat.[77] The reflectors are simple 50-mm-diameter plastic spheres, which give the animals a strong echo from their echolocation calls.

Recent experiments with this system (box boxes and reflectors) revealed more details about how bats might find suitable roosts.[78] Specifically, in the field bats used associative learning, spatial memory and social information to distinguish between roosts. This combination of cognitive processes further extends our knowledge of the dynamics of bats. It is no surprise that bats make the best use of available information to locate and perhaps evaluate features as important as roosts. Having this information is of great potential value to those working on bat conservation.

Soprano Pipistrelles flying around a bat house. The bat house is erected on posts about 4 m above the ground. Arrows indicate entrances to roosting slots between the inside partitions. JR

9 Social lives of bats

Many bats are gregarious, often roosting in groups that range in size from tens to millions of individuals. This, by itself, does not make them 'social' animals. The basic social unit in mammals is a female and her dependent young. For bats, this means that two is the usual minimum group size (a female bat and her young). In Eastern Red Bats, the unit can be three or four, which is exceptional among bats. As in most mammals, during the mating season males join females, sometimes for brief encounters (mating), other times for longer periods.

On Christmas Day 2017, this female Wahlberg's Epauletted Fruit Bat and her young roosted under the eaves of The Store at Skukuza (Kruger National Park). Mother–young pairs of these bats roosted under the thatch roofs all around the building. They were nicely out of reach of passing tourists, most of whom were completely unaware of the bats. The pup is smaller than its mother, and slightly lighter in colour.

Opposite: Pallas's Long-tongued Bat and a pup roosting under a thatched roof.

This mother Eastern Red Bat has her wings wrapped to mostly cover her three small young. The young are naked and pink in colour. The forearm of one is obvious and much smaller than those of its mother.

Reproduction

Bats have an astonishingly low rate of reproduction. Females of most species bear one young at a time, although a few have twins. Most species have one reproductive season a year. Gestation periods of bats are long compared to other mammals of similar size, and the young are, in relative terms, huge at birth (up to about 30% of a mother's mass). This is the equivalent of a 68 kg (150 lb) human mother having a 20 kg (45 lb) baby. While nursing, a dependant pup consumes its own weight in milk every day, arguably a much higher investment than that associated with pregnancy. For most of the 1,400+ species of bats, we have few details about the interactions between males and females around mating.

Bats are sexually dimorphic, so males and females look different. The simplest distinction is the penis of males, which is of course lacking in females. Reproducing females produce milk from a pair of pectoral mammary glands, and even outside of periods of lactation, the nipples are distinctive (see page 12). Some male Dyak Fruit Bats from Malaysia and Borneo also appear to produce milk, albeit in much smaller quantities than females. Male lactation also occurs rarely in other species of mammals, including humans. It is not clear if the record of male lactation in Dyak Fruit Bats was related to some hormonal dysfunction or to a normal situation.

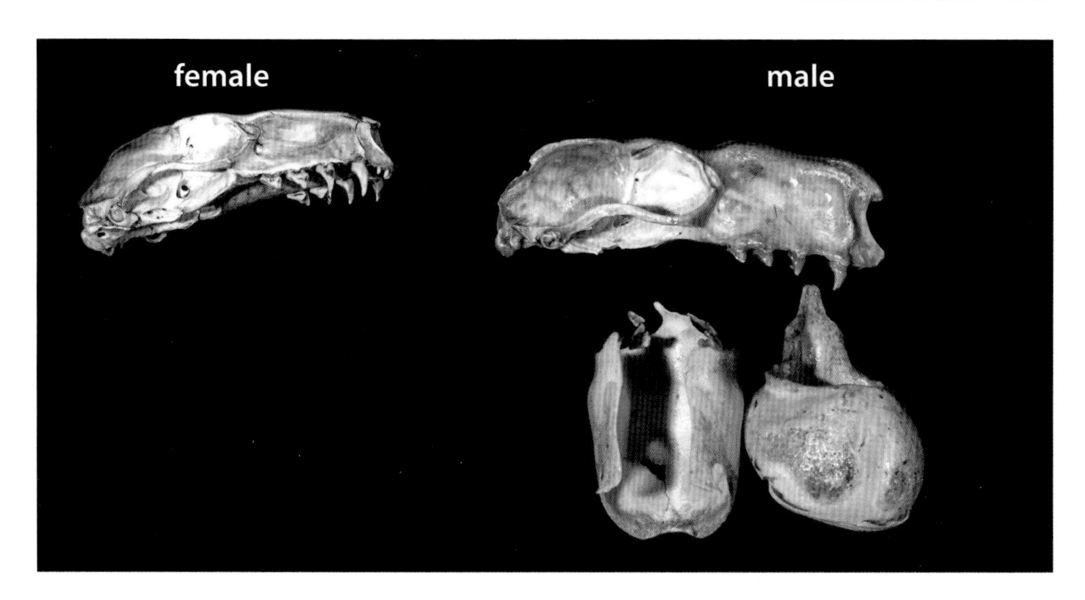

Sexual dimorphism is striking in Hammer-headed Bats (>300 g). Males are larger in overall body size, and notably in skull size. Also shown are the components of the male's resonance chamber (enlarged larynx, thyroid and cricoid cartilages). Males call to attract females.

Some species of bats are also sexually dimorphic in terms of their behaviour, body size and various glands. In Hammer-headed Bats, males gather and call to attract females. These aggregations of males are called 'leks', the sites where mating occurs.[79] Females come to the lek to mate, having a choice of several males right at hand. The skulls of males and females in this species are distinctly different, coinciding with differences in calling behaviour. At some sites most visiting females choose the same male.

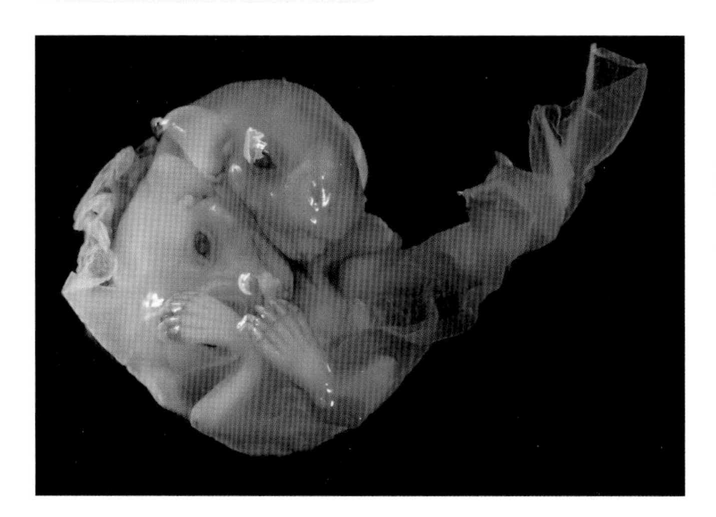

Twin bats may be produced from one egg or two. There are two records of conjoined twins in bats: Pygmy Fruit-eating Bats from Belize (pictured left), and Big Brown Bats from Canada.

As in many mammals, male bats contribute sperm to reproduction, but not much else. In species such as Little Brown Myotis, both males and females are promiscuous, mating repeatedly throughout the season. This has interesting implications for mate choice. Female Little Brown Myotis bear one young a year, but probably each mates with more than one male. How does a female choose the sperm that fertilizes her egg or who is the father? Big Brown Bats in eastern North America bear twins. Genetic studies revealed that 60% of twins have different fathers. Again, a challenge is to find out how paternity is determined. Mother bats invest heavily in their young, suggesting that females should choose the best available male to father her young. Litters with mixed paternity may be the result of females hedging their bets. Conjoined twins occur rarely in bats.[80]

A mother Silver-tipped Myotis (7 g) with her very young pup. The pup is almost hairless, but its eyes are open. Note the very short and stubby forearm (compare with the photograph on page 148).

Two young Big Brown Bats, upper aged 3 days, lower aged 11 days.

In most species of bats, mating is followed directly by fertilization, pregnancy and birth. In the tropics, the mating season is timed so that young are born when food is plentiful, initially so the mother can produce milk for her offspring, and then so that when young are learning to forage there is an abundance of food.

Bats also show two other patterns, delaying fertilization or development. In the temperate region, females of many species of bats store sperm. Big Brown Bats and Northern Bats, for instance, mate in the autumn. Females store sperm in their uterus from mating until they leave hibernation in spring. In some species, this may be over 200 days. This strategy means that most mating activity occurs in late summer and early autumn while bats still have access to abundant food, before colder temperatures stop most flying insect activity. Then females get pregnant in spring and the young are born in early summer. Young bats grow quite quickly, as illustrated by comparing a three-day-old Big Brown Bat with a much younger Silver-tipped Myotis or Eastern Red Bat (see pages 15 and 148, respectively), let alone an 11-day-old Big Brown Bat.

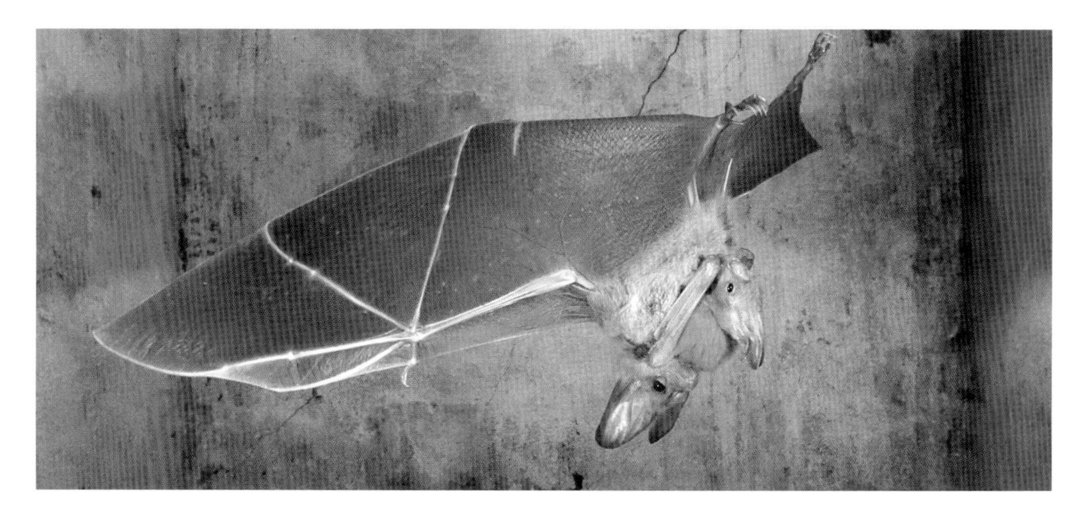

Two mother bats carrying young. The Heart-nosed Bat (25 g) emerging from her roost in a well (above), while the Lesser Short-nosed Fruit Bat (30 g) carries her young as she visits a flower (below). *JR*

Biologists often catch mother bats carrying their young. Sometimes mothers are moving their young from one roost to another. In Africa we have watched Egyptian Slit-faced Bats doing this, perhaps to leave the young in a 'safer' temporary roost while she forages. Fruit-eating bats sometimes carry their young when foraging. Photographing bats emerging from roosts often reveals females carrying their young. On occasion, mothers carrying young suggests a response to disturbance. Recent work with Egyptian Fruit Bats has revealed that

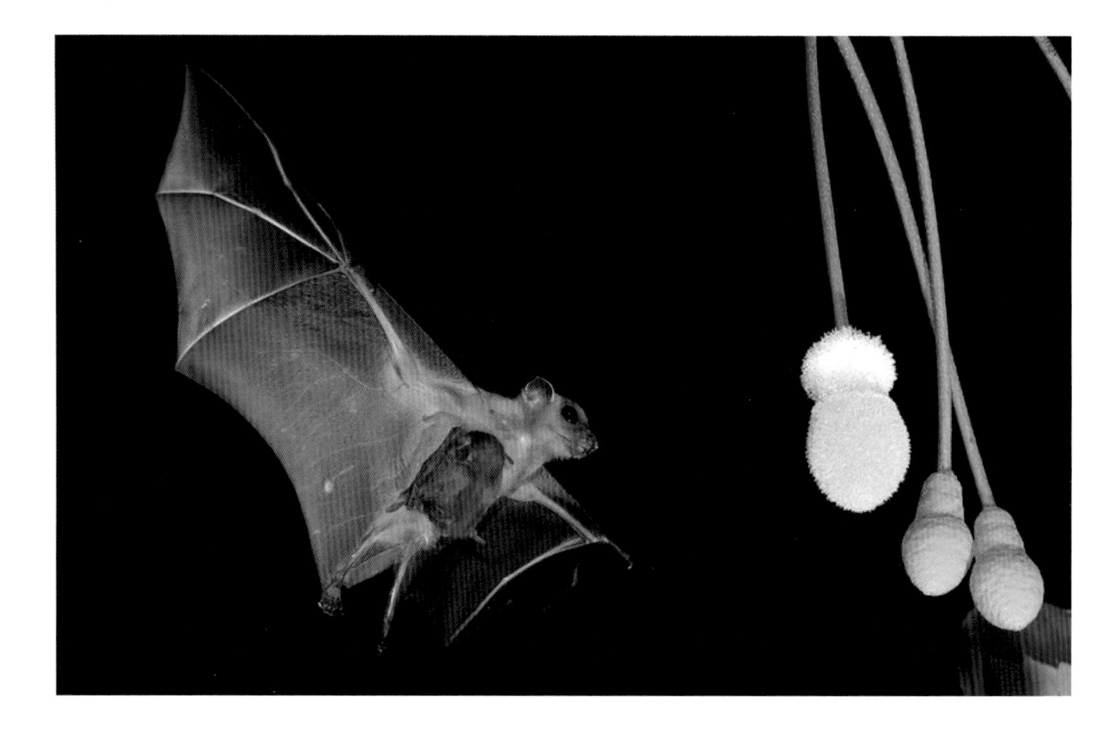

pups carried by their mothers actively learn about their home ranges.[81] These studies extend our knowledge of how bats can learn about their surroundings.

In Seba's Short-tailed Fruit Bats and California Leaf-nosed Bats, fertilization immediately follows mating. But these bats delay processes of development long enough to ensure that young are born when food is plentiful. Female Seba's Short-tailed Fruit Bats can arrest development at a very early stage of development, the 'primitive streak', for up to 60 days. This can be induced in the laboratory by reducing food availability.

We know relatively little about the weaning process in bats. Although there are records of female bats and their young flying together, it is not clear if weaning, the change from a milk diet to one of insects or fruit, fish or blood entails females bringing food to their young. Little Brown Myotis usually are flying and echolocating by the time they reach 21 days, Big Brown Bats by about 28 days. The double challenge of learning to fly and to echolocate probably requires females to continue nursing and to deliver insect food one way or another.

Female Yellow-winged Bats in the savannah in East Africa have one young each year. Young cling continuously to their mother from birth, usually attached to an inguinal nipple (in the groin area); they then nursed from pectoral nipples. Researchers observed that as the pup grew, it hung by its hind feet from its mother's shoulders and flapped its wings.[82] This process appeared to underly the development of flight. Young slowly became proficient flyers and accompanied their foraging mothers, following about 10 m behind. Young were weaned by 20 days after their first flights. After that, for at least 50 days the young continued to roost with their mothers. Details such as these are lacking for most species of bats.

Yellow-winged Bat. *JR*

What is a colony of bats?

Some bats live in social units which may be called 'colonies', but what is a colony? Individual social bats recognize one another by scent and/or sounds. We know most about the social situation in Common Vampire Bats, probably the most social of all bat species. Work on these bats in Costa Rica revealed that, on any given night, individuals in a colony might not roost together. But over a longer time they are 'social' because they recognize one another and occasionally roost together. Interactions among group members include grooming one another. Bats use vocalizations and/or odours as keys to recognition. In Common Vampire Bats, tight social bonds extend to food-sharing among members of a group. Yet finding bats together in a roost (like people at a bus stop) does not mean that they are a social unit. The social unit might be much larger than the individuals roosting together on any one night, or there may be distinct groups of individuals within larger populations in a roost.

A male Greater Sac-winged Bat (8 g) showing wing sacs (arrows), one enlarged on the animal's right wing. The bat is emerging from its roost in tunnel in Belize. These bats roost in areas with considerable amounts of light, for example simply under the eaves of buildings or between buttressed tree roots. Opposite: close-up of one of the sacs.

In many tropical bats, reproductive groups consist of a male with several females and their dependant young (sired by the male). Males defend the roost used by the group, often welcoming other females, but not other adult males. Advertising the group's territory can involve different displays. Adult male Greater Sac-winged Bats advertise their territories by flight displays, songs and odour. Sacs in the male's wings are fermentation chambers in which males deposit spit, urine and other secretions to give off their pungent individual scent.

Male Northern Yellow-shouldered Bats (see page 97) have shoulder glands that produce an aromatic secretion used to attract females.[83] Analysis revealed 15 compounds in extracts of fur around the shoulder gland – three of them occurred only in adult males, not in females nor subadults. As in some other mammals, males may mark females, young and surfaces with glandular odours. Such marks may attract adult females and repel other males.

The development of proximity tags allowed researchers to show that Common Vampire Bats had 'friends'. When two individuals, each with a proximity tag, come close together the contact is recorded and available to the researchers. They spent more time with individuals that they could receive blood from, or donate blood to, as the need arose. Put in context of previous and ongoing interactions, researchers accurately identify the bats 'knew' one another. Strangers were unlikely to successfully beg blood from group members.

In many species of bats, the social structure underlying group composition is known as a 'fission–fusion society'. This describes long-term social organization of subgroups within larger groups. In bats, some roost-mates may be genetically related, but the social bonds extend beyond blood lines. In Common Vampire Bats, individual recognition and blood-sharing are cornerstones of social organization. We should recall the importance of individual recognition and how it affects our own networks.

Food availability and social patterns

In 1978, a Canadian researcher working in the Chiricahua Mountains of Arizona studied bats' use of rich patches of insect prey.[84] He used battery-operated ultraviolet lights to attract insects and a bat-detecting system to monitor bats' responses to the patches of food he created. The bat detector allowed him to identify species by their echolocation calls and use feeding buzzes to recognize foraging-event behaviour. He found that while some species of bats readily exploited the concentrations of insects, hunting and catching their prey right there, others rarely if ever came to the patches. This was an effective approach for the time because it did not involve capturing and marking the bats.

More recently, the advent of tags that provide GPS information, as well as recordings of what bats said and heard, revealed two interesting patterns of foraging behaviour. Foraging bats that searched for ephemeral prey (Fishing Myotis and Greater Mouse-tailed Bats) regularly changed foraging areas, often foraged at different times of the night, and aggregated with conspecifics after localizing a concentration of prey. Meanwhile, species that hunted for more predictable food sources (Greater Mouse-tailed Bats, Egyptian Fruit Bats and Lesser Long-nosed Bats) showed high fidelity to foraging sites and never foraged near conspecifics.[85] Playback of echolocation signals attracted the ephemeral foragers but not those searching for predictable sources of food.

Female Lesser Long-nosed Bats congregate in underground roosts, from where they forage widely for nectar and pollen, mainly obtained from Saguaro cacti (see pages 104–5). These bats appear to visit cacti in an orderly sequence, creating relatively small non-overlapping cores of cactus plants. Each bat needs 14–19 ml of nectar per night but can only obtain 0.65 ml from any one flower.[86] During migration, Lesser Long-nosed Bats appear to move and forage in groups. This species and Mexican Long-tongued Bats regularly visit and drain hummingbird feeders left out overnight. They show no obvious territorial behaviour in these situations.

Lesser Long-nosed Bat in southern Arizona

There are significant advantages to tags that provide information about movements, locations, vocalizations and other signals of known individuals. They can allow assessment of interactions among foraging individuals even when they are not visible.

Greater Horsehoe Bat adults and young in a roost.

Observational learning

Working with captive bats can be very educational. In the early 1980s, a graduate student quickly noticed that some of our Pallid Bats (15 g) paid close attention to others eating mealworms.[87] She soon realized that once one of the Pallid Bats learned to fly to a target with mealworms, the others soon followed suit. Astutely, she repeated the experiment with both Big Brown Bats and Little Brown Myotis. An inexperienced bat placed in the room with the target and the mealworms did not learn to associate target and food. Furthermore, any of the three species could be a model (tutor) for others to follow, which meant interactions

Pallid Bats in a night roost close to the Rio Grande River in Big Bend National Park in Texas. Bats use night roosts as places to rest and digest after feeding. Groups of bats in night roosts also may exchange information, if only through observational learning.

within and among species. The bats all learned by observing others. An individual typically learned the novel feeding behaviour in a 25 minute session with ten trials.

You will recognize this as a variation on Fringe-lipped Bats (page 72) learning to associate food with a Bob Marley song. Capacity for learning by observation is likely widespread among bats, especially species that are long-lived and live in social groups. This also raises questions about how young bats first learn to fly, echolocate, find and catch their own food.

10 How bats use space

Many aspects of bat life history and behaviour hinge on how these animals use space and habitat. They must have roosts and places to forage. Reproductive condition as well as changes in details of weather or season can influence what they do when. But if you are trying to conserve/protect bats, how big an area does a colony of Big Brown Bats or Northern Bats, for instance, require to survive and thrive? Are the areas the same in summer and in winter? What about much larger or much smaller species? This chapter illustrates how technology and innovation have dramatically increased our knowledge of how bats use space.

Banding studies in the 1950s and 1960s revealed that Little Brown Myotis banded in summer were rarely recovered in winter, and vice versa. A 1965 paper reported recovery of just 18 of several thousand banded individuals at a site other than where they had been banded.[88] In 1966 MBF was overjoyed to find a female Little Brown Myotis in an attic about 200 km from where she had been banded in winter in a mine at the north end of Lake Champlain, New York State. From 1965 to 1969, MBF spent hours and hours in tens of different roosts, banding several thousand bats. His results did not change the picture. Wherever/whenever the research was conducted, the pattern remained the same: simply, there are many, many bats and roosts/hibernacula out there and we are only ever able to access a small proportion of them.

Homing studies have shed some light on how bats use space. Little Brown Myotis (and other species) were banded in their roosts, taken some distance away and released.

When this PIT-tagged Fringe-lipped Bat flies through a loop antennae (opposite), a reader records information about the bat: the time(s) it appears and any other movements it makes through the loop. A Tricoloured Bat (6 g) hibernating in an abandoned mine in south central Ontario (right). Note the condensation in the hibernating bat's fur. Tricoloured Bats' lives are relatively unknown. Until about 20 years ago we presumed they roosted in hollows, but radio-tagging revealed that they roosted in foliage.

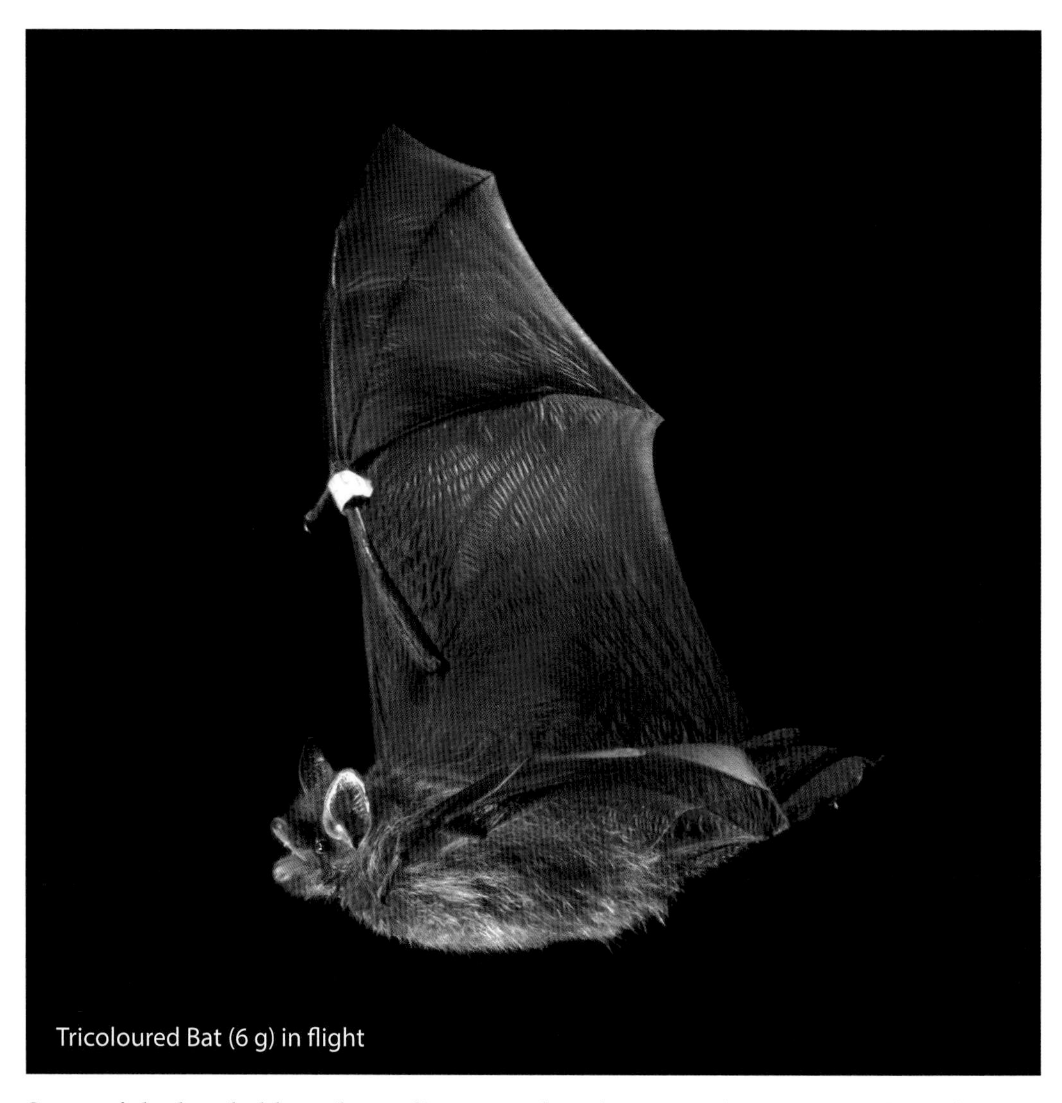

Tricoloured Bat (6 g) in flight

Some of the banded bats 'homed', returned to the original roost, but others did not. As with band recoveries in a hibernation site, you know the ones you recovered but unrecovered bats could be dead, there but not findable or in some other site not known to you. This is a reality of passive tags such as bands. Active tags send a signal, allowing the researcher to find animals within range of the tag's signal and the receiver.

Changing to active tags gives more opportunities to find and follow bats. When radio transmitters weighing about 0.5 g became available (see page 16), we used them on Little Brown Myotis. We found that individuals moved up to 5 km each way going to and from foraging areas. We also learned that Little Brown Myotis often fly continuously for several hours a night. This information shed some light on the bigger question of how Little Brown Myotis might use space, because we now knew that these bats could cover longer distances when necessary. Four hours flying at 5 $m.s^{-1}$ (18 $km.h^{-1}$) could be over 70 km. But the ranges of the transmitters were 1 or 2 km, as limited as the picture they provided.

In 1967, colleagues used radio tracking to study homing Greater Spear-nosed Bats (84 g) in Trinidad.[89] They caught bats in their cave roosts, attached 7-g transmitters to them and then displaced them several kilometres from their roost. Some bats were blindfolded, others wore goggles, still others (control) were unencumbered other than by the transmitters. Signals from the transmitters allowed documentation of the routes the bats took home. All the bats made it home, the bats with goggles almost as quickly as control animals. Blindfolded animals took longer and used more circuitous routes. In addition to demonstrating that the animals knew the area, it also suggested that vision was part of recognizing features of the landscape. Blindfolded bats could also have been more vulnerable to predators.

At a next level, in Malaysia, researchers outfitted 700-g male Large Flying Foxes with satellite tags (either 20 g or 12 g).[90] As the name implies, signals from the tags relayed to a satellite allowed researchers to determine the bats' flight paths. These bats travelled hundreds of km between roosting sites, frequenting roosts and foraging areas well beyond Malaysia and extending into Indonesia and Thailand.

Technology has advanced, allowing us to obtain more details about how bats use space. The satellite tags used on flying foxes in Malaysia are themselves heavier than many species of bats. By attaching weights to bats and flying them through obstacle courses, researchers showed that when a tag exceeds 5% of the animal's body mass, it negatively affects their manoeuvrability. This restricts our choices of species to study with active tags, but the scene is ever changing as lighter and lighter active tags become available.

Two types of tags and systems deserve special mention, PIT (passive integrated transponder) and Motus (see pages 15–16). Both have been used on Little Brown Myotis (as well as many other species) with considerable success. The results reveal scales of movement not previously considered likely or possible. In 2009, a reviewer of a paper about Little Brown Myotis on Long Point, Ontario, bridled at the suggestion that these bats could fly the 34 km from the tip of Long Point to the south shore of Lake Erie, which lies in the United States. We already knew from radio-tracking studies that 35 km was quite feasible for Little Brown Myotis. PIT tags are small glass tubes containing an electromagnetic coil with a single memory chip that holds an individual code, and a transmitter. The smallest ones have been used on Little Brown Myotis. A PIT tag is subcutaneously injected into the animal and can be read when the tag passes close to a receiver. People use these tags on their dogs and cats to facilitate reuniting with lost pets. Since at least 1989, PIT tags have been used to keep track of individual bats of various species. Working in Newfoundland, researchers documented moves of tagged Little Brown Myotis between roosts (bat houses) 35 km apart.

PIT tag readers installed at the entrances to bat roosts lead to interesting discoveries. For example, a Lesser Long-nosed Bat with a PIT tag travelled at least 600 km from where it had been tagged (Isla Carmen in Baja California Sur) to a roost in southeastern Arizona. The individual was tagged in May 2015, then detected in Arizona in September 2020. The researchers learned about the distances individuals could travel, as well as about how long the tags can be active. A huge advantage of PIT and Motus tags is the ability to collect information about tagged bats without having someone trying to catch them.

Motus tags detect motion, allowing biologists to track even very small bats, as well as almost any other animal you can imagine. Small Motus tags (nanos) are just 0.2 g to 2.6 g in weight and produce uniquely coded amplitude modulations of a coded signal, which is detected by a receiver when a tag is within range. These tags have also been used successfully on Little Brown Myotis, revealing that some tagged individuals flew 100 km nonstop in one night. A Motus nano tag weighing 0.4 g has a 19-cm-long antenna (see page 15) but is small enough for a Little Brown Myotis.

The success of Motus tags hinges not only on tagged individuals, but also a network of receiving stations to record passing bats (or birds or insects).[91] In the case of Little Brown Myotis, a tagged bat moved the 100 km between stations, but we do not know where it started from or where it was going. Along Long Point, a network of stations currently cover an area 20 by 40 km, allowing documentation of the movements of 30 tagged Silver-haired Bats. Before this work, evidence of their migrations was negative – in that we presumed these bats went south because we never found them in southern Ontario from late September to early May. Data collected from bats with Motus tags showed two waves of these bats moved south through Long Point, one in late August, one in mid-September. Half of the tagged bats flew south from Long Point across Lake Erie. The rest followed the shoreline around the lake before eventually going south. Based on distances bats travelled between stations, they were covering 250–275 km per day. The tagged bats spent some time in the Long Point area, feeding at night, and using torpor during the day.

Motus tags have also been used to advantage in documenting how bats use habitat. Researchers at the Toronto Zoo caught Northern Long-eared Bats (see page 14) in a forested area at the northeast corner of Toronto.[92] The forest is surrounded by urban sprawl and is bisected by a major six-lane highway (401). Tagged bats roosted in several different trees and foraged in the forest. At least one crossed the highway (probably by going under a bridge) and foraged in a large marsh. We still do not know where these endangered bats overwinter.

Some bats migrate, although the evidence for

A Natal Long-fingered Bat emerging from a cave in Limpopo Province of South Africa. A PIT tag reader at the cave entrance documents the comings and goings of tagged individuals. Photo courtesy of Ernest Seamark.

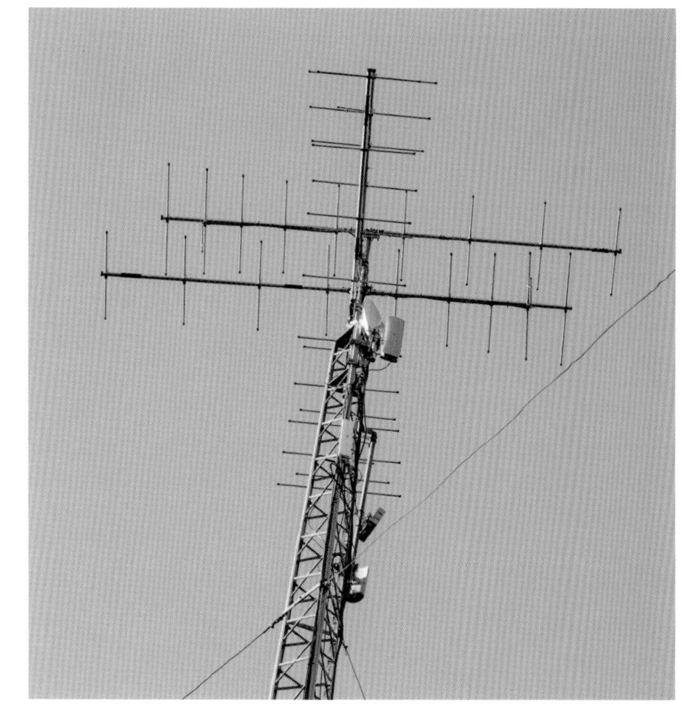

A 30 m tall Motus tower set up at the base of Long Point (Ontario), a through area for migrating bats and birds. The system detects tagged animals, allowing us to document the movements of individuals.

this in many cases is only presence/absence. Many species are present in some areas for part of the year, and then gone from there for the rest of the year. The list is long and includes bent-winged bat species (10 g) in Africa, Wrinkle-faced Bats (20 g) in parts of Central and South America as well as Silver-haired Bats (10 g), Hoary Bats (30 g) and Eastern Red Bats (12 g) in North America. Brazilian Free-tailed Bats (10 g) migrate from parts of South America north to Colorado in the United States. Scattered band recoveries had supported the idea of migration. Other species are more sedentary. Big Brown Bats with Motus tags rarely move more than 12 km from where they had been tagged. Here the data from Motus tags mirror many recorded movements of banded individuals.

Some species clearly have considerable dispersal capability. But we still have limited knowledge about a bat's home range, even for well-studied species. It would be great to know how a bat learns its home range and the locations of essential resources such as roosts and areas to forage. Do bats (some or all) have cognitive maps, internal neural representations allowing an individual to register, store and recall information about locations relative to their survival?

Until recently, only humans and a few other animals showed evidence of using cognitive maps to navigate their home ranges. Two eye-opening reports in 2020 demonstrated that Egyptian Fruit Bats (see pages 5–6) had cognitive maps they used in the same way as certain other animals.[93] The research drew on data collected by sophisticated tags that provided details of where the bats went and when. Two separate groups of researchers in Israel collected the data. They documented bats' use of space – how they move efficiently and directly to specific goals, such as fruit trees and roosts. By tagging pups before they could fly,[94] the researchers learned how young progressed from flying around in their roost, to going farther and farther afield, taking about 60 days to learn their home ranges, which were often at least 60 km^2. Further work revealed that spatial and temporal information was stored in the bats' hippocampi, an area of the brain.[95]

Greater Mouse-eared Bat (40 g) at a swarming site (Nietoperek) in Poland. This military bunker was built by the Nazis and now is used for hibernation by tens of thousands of bats. Those that hibernate here also visit during swarming. *JR*

A Silver-haired Bat (10 g) with Motus tag photographed at Long Point, Ontario.

In the northern hemisphere, August and September are interesting times to study the movements of bats that hibernate underground. If you visit a hibernation site (cave or abandoned mine) during the day in August, you rarely find any bats. But if you return to the same site an hour after dark, you find many (tens, hundreds) flying in the passages. This behaviour, known as 'swarming' is well documented. The swarming bats typically include the species known to hibernate at the site in question. Banding studies revealed that different individuals visit the sites from night to night. Interestingly, if you caught and banded bats on a date in August in one year, you often recapture some of the banded individuals on the same date, year after year.[96]

The swarms include adults and subadults, males and females. Swarming could be a way for young animals to learn the locations of hibernation sites. It also is part of mating behaviour and, in September, the incidence of mating bats increases. One Little Brown Myotis originally tagged in northern Ontario in winter, was then caught swarming at a mine near Renfrew, Ontario (see page 15). In Europe in August, Common Pipistrelles often visit buildings they use as roosts – a possible equivalent to swarming at underground hibernation sites.

Swarming behaviour provides photographers with excellent (but challenging) opportunities to photograph flying bats. You usually can identify the species from the photographs and the approach is much less disruptive to the animals than capture and handling. Camera trapping also can be used to advantage at other roosts.

So, bats' use of space is much more complicated than we had expected. The evidence comes from PIT tags and Motus tags, combined with data about cognitive maps. These findings help to put data (both on bat survival and dispersal) from banded individuals into broader context. Studies of stable isotopes also reveal patterns of movements of bats.

Two Common Vampire Bats emerging from a roost in a hollow tree. Note that the lead bat is banded. In this study area, although most banding of these bats has been done at one site, photographs from other locations reveal movements between roosts.

Bats get around

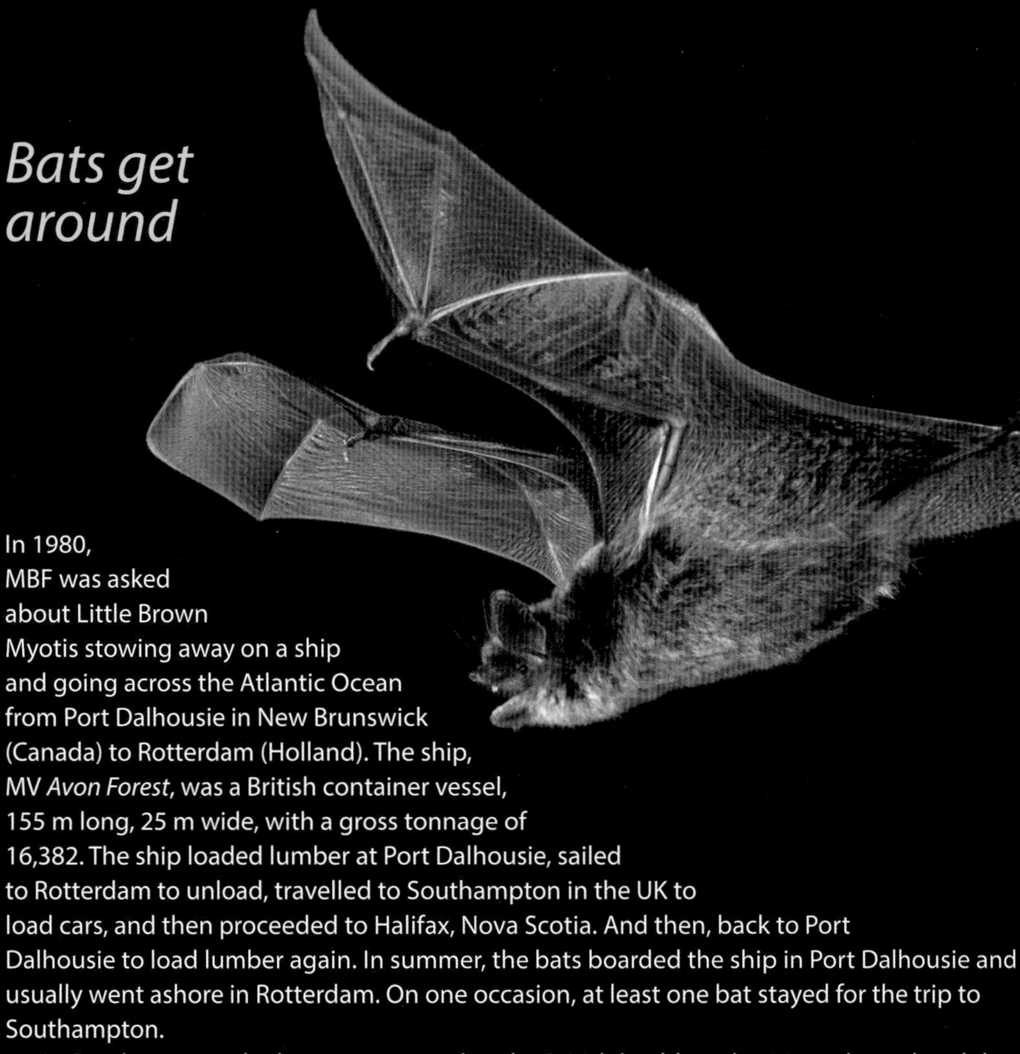

In 1980, MBF was asked about Little Brown Myotis stowing away on a ship and going across the Atlantic Ocean from Port Dalhousie in New Brunswick (Canada) to Rotterdam (Holland). The ship, MV *Avon Forest*, was a British container vessel, 155 m long, 25 m wide, with a gross tonnage of 16,382. The ship loaded lumber at Port Dalhousie, sailed to Rotterdam to unload, travelled to Southampton in the UK to load cars, and then proceeded to Halifax, Nova Scotia. And then, back to Port Dalhousie to load lumber again. In summer, the bats boarded the ship in Port Dalhousie and usually went ashore in Rotterdam. On one occasion, at least one bat stayed for the trip to Southampton.

In Southampton, the bat was reported to the British health authorities, who ordered the ship fumigated. MBF was asked what could be done to keep this from happening again. A visit to Port Dalhousie revealed the size of the problem. The ship was huge and the cargo space vast. But every evening Little Brown Myotis foraged around the docked ship, taking insects from the water and, after filling up, used the hold as night roost, a place to digest their meals. Over the five days it took the ship to load, the bats became accustomed to the roost. The ship usually sailed at midnight, with the stowaways safely on board.

It was said that everyone in the crew knew about the bats, and at least one had been kept as a pet. But, back to the question, how to keep the bats from doing this again? The bats hibernated elsewhere so it was a summer problem. Sailing at noon seemed the easiest solution.

11 Threats to bats

Without doubt, continued exponential growth of the human population and its attendant impacts on habitats is the major threat to the survival of bats and very many other species. The survival and continued existence of bats depend on roosts and food. Expanding human populations generate demand for more agricultural land, which can mean less suitable habitat for bats. Urban areas are the fastest growing ecosystems in the world, their growth coinciding with loss of prime habitat that is home to many organisms. Bats' extensive use of human structures as roosts shows that some species persist and thrive in urban settings, but others do not.

Bats can have a positive economic impact. Tourists' visits to sites with huge concentrations of bats at caves (e.g. Bracken Cave near San Antonio, Texas) or under bridges (e.g. Congress Avenue Bridge in Austin, Texas) can have a significant positive effect on local economies (see page 196). Bats' voracious appetites mean that they can consume vast numbers of insects. Furthermore, bats' roles as pollinators of important plants, and as seed dispersers, means that they play an important part in ecosystems.

Small-scale forest clearing for agricultural land in Belize. Note crops in the upper part of the picture, and three small patches of remnant forest.

Predators

The rate at which they encounter prey may be as important to predators as the time it takes them to process a capture. The combination of bats' small sizes and the tendency for many species to live in large groups makes them attractive to a range of predators. For a predator, clouds of bats emerging from a roost can mean very high rates of encounter and, because of their small size, short processing times. No wonder predators often congregate around roosts with large bat populations. If you visit Bracken Cave, for example, look for the various species of predatory birds that harvest bats from the emerging swarms. This is a recurring theme at bat roosts with large populations in many parts of the world. The range of predators includes birds of prey, snakes, as well as mammals, from martens to house cats and raccoons. The list even includes some spiders.

In March 1992 a group of us studied bats and raptors in Kruger National Park in South Africa.[97] There, three species of raptors commonly perched outside bats' roosts just before the bats began to emerge. The birds were typically in place by about 40 minutes before dark. Bats roosted in large numbers in buildings and under bridges. Emerging bats presented the birds – Wahlberg's Eagles, African Goshawks and Hobby Falcons – with an important opportunity. Individuals of these three species succeeded in about half of 59 attacks on bats, whether they dived on one bat or just chased it down. These birds always took a captured bat to a perch and ate it there, taking between 30 and 300 seconds from capture to resumption of hunting. By hand-releasing some bats we had caught, we tested the idea that bat emergence stimulated the birds to take off to be ready for bats. Sure enough, the birds took off in response to even a few bats flying near the roost.

A Cuban Boa photographed in a cave many bats use as a roost. This is one of a group of bat-eating boa species found in the Caribbean. These boas sit near entrances to the cave, or in narrow passages which funnel flying bats and put them within striking distance.

Raptors need to eat 15–30% of their body mass a day. For a Wahlberg's Eagle, six 20 g Angolan Free-tailed Bats in less than 15 min would be 80 g, or 40% of its daily food intake. We noticed that colony size influenced time of emergence. Colonies of fewer than 100 bats had somewhat haphazard emergence times, while those with more than 100 bats tended to emerge earlier. Emerging before dark exposed the bats to the raptors, but the situation was fluid because when a raptor was busy eating a bat, it could not be hunting for another one. At larger bat colonies, handling time (from capture to ready to hunt again) was the main determinant of capture rate for Hobby Falcons, African Goshawks and Wahlberg's Eagles. The bottom line is that large bat colonies can be important food sources for these predatory birds and represent a significant part of their overall intake. Many other observations indicate that raptors and other birds can be more opportunistic predators. Other birds such as jays, crows, ravens, magpies and shrikes can be adept at catching day-flying bats, which they encounter more by chance. The Gymnogene, or African Harrier-hawk, hunts by reaching into hollows in trees to extricate and eat roosting bats (and other animals). Bats are of course largely nocturnal, perhaps reflecting an adaption to their vulnerability to predatory birds.

Swarms of insects offer bats the same opportunity that bats offer raptors. Swarms of insects mean high rates of encounter and short handling times. Insects are often very numerous in the hours just before and after dark when there is a mixture of flying diurnal and nocturnal species. Yet bats that emerge and begin to feed before dark must cope with birds that hunt then.[98] The Bat Hawk, for example, is more specialized and highly dangerous for bats. These hawks catch flying bats, swallow them whole in flight, and within 6 seconds are foraging again! Bat Hawks occur from Africa to Southeast Asia. They typically hunt swifts and bats emerging from or returning to roosts. Many other birds of prey opportunistically hunt bats at dusk. Most dangerous are small falcons and hawks which often patrol near bat colonies during emergence and return.

Levels of lighting directly influence some bats' risk of predation. This could explain changes in bats' behaviour during the full moon. Predators such as night monkeys or house cats successfully catch flying bats when there is enough light. Indeed, at building and cave roosts, cats appear to watch for emerging bats, sometimes jumping and grabbing bats as they fly by. At some roosts in Belize, Common Vampire Bats do not emerge until the moon has gone down.

Owls are sit-and-wait predators, and often scan from a vantage point near bat roosts, catching bats during emergence and return. In Europe, Tawny Owls and Barn Owls are examples of species that kill many bats.

In light of the tradition 'it takes a thief to catch a thief', it is not surprising that some bats eat other bats. Most of the details of this come from occasional observations, for instance regarding the Large Slit-faced Bat (see page 82). Spectral Bats have been known to take other bats from mist nets. Discarded remains below the roosts of Australian Ghost Bats and false vampire bats also indicate that bats can be prey for others of their kind. More recently in Puerto Rico, biologists discovered that Greater Bulldog Bats (65 g) eat smaller species that roost in the same caves, namely Antillean Fruit-eating Bats (46 g) and the much smaller insectivorous species Sooty Moustached Bats (6 g). The Greater Bulldog Bats caught and ate unweaned, young Antillean Fruit-eating Bats about the same size as the adult Sooty Moustached Bats. We do not consider as records of predation reports of larger bats eating smaller ones when confined together in a holding cage or bat trap. With more data, it will be interesting to see the role bats play in the diets of other bats.

House cats are ambush predators, often lurking around bat emergence sites. Feral cats living in and around caves in the Caribbean – in Cuba, Jamaica, Haiti, Hispaniola, Puerto Rico, for example – take a large toll on bat populations, including some that are endangered species. In caves, hunting stations used by such cats are often surrounded by piles of discarded bat wings. In Jamaica, a mother cat and her kittens may kill and eat two or three bats a night. In many of the same caves, resident boas also hunt and catch emerging bats. These cold-blooded predators do not have the same energy requirement as warm-blooded cats and birds, and so their impact on bat populations may be less.

House cats, domestic or feral, pose a considerable threat to bats, and could be important in transmitting diseases from bats to people. Cats living in houses with people infected by COVID-19 (see page 210) show evidence of exposure to SARS-CoV-2. This means that COVID-19 could be transmitted from people to house cats, but also to much larger cats, such as tigers in a zoo. In parts of South America, there are records of cats with the vampire bat strain of rabies. This could be due to two possibilities. A cat could have been exposed when a vampire bat, perhaps unwisely, decided to drink its blood. But a cat could have been exposed when it caught and ate vampire bats. The two possibilities are not mutually exclusive. Worldwide, dogs are a major vector for the transfer of rabies to humans (see below).

Often, bats avoid predators by waiting in the roost until after dark. Another defence is to swamp predators by emerging en masse, a strategy that can work for individuals while there will always be a certain amount of losses among the wider group. In some tropical species, white or transparent wings may camouflage bats against the night sky (see page 32). Bats appear to select roosts with great care, particularly if the roost will be a maternity site. Here, a good roost must be warm and minimize the risk of predation.

Many predators take bats opportunistically. In Costa Rica, some squirrel monkeys were quite effective at catching bats roosting in their leaf tents.[99] The monkeys clearly recognized the tents and then, from the ground, checked them for bats. Having found bats in a tent, a monkey would climb into a tree overhanging the tent, jump on the tent then quickly grab and kill the fleeing bats. Male squirrel monkeys were more effective at this than other group members. In Jamaica, a large (7 cm long) female Golden Silk Spider caught a Velvety Free-tailed Bat (see page 12)

Hunting station used by a feral cat in St Clair Cave in Jamaica. Note the discarded, mouldy bat wings.

A female Golden Silk Spider (body length 40 mm) on its web in which it caught bats.

in a web just below the entrance to an attic roost. The flexibility of opportunistic predators, including bats, can be astonishing.

Hibernating bats appear to be defenceless, making securing a safe roost a matter of life and death. Here, a safe roost must be cool but above freezing, dark and out of the reach of predators. This means hibernating bats roost in caves, on the ceilings, in cracks and crevices, or deep among scree on the floor. Mammals that prey on hibernating bats include rats, martens, raccoons, cats and even deermice (*Peromyscus* sp.). Astonishingly, at some sites in Europe, Great Tits have been observed flying into caves to grab hibernating Common Pipistrelles.[100]

When first disturbed, a hibernating bat opens its mouth wide and produces an intense hissing sound. In a laboratory setting, deer mice would not closely approach a bat producing this display. The sound could be intense enough to be painful, which would explain the mice's reluctance to approach the bat. The behaviour appears widespread among bats that hibernate.

People also prey on bats, and human predation accounts for dramatic reduction in populations of some flying fox species on islands in the South Pacific. Traditional hunting of these bats involved snaring of roosting individuals in nooses on long poles. The advent of shotguns allowed more rapid bat harvesting. In many parts of the world, bats are 'bush meat' and hunted to be used as an important source of protein.

Mishaps

Accidents can also kill bats. For example, there are records of bats dying in narrow-necked vases where the slippery surface prevented bats from climbing out, and the neck of the vase was too narrow for them to fly out. In these situations, for instance in an attic, people may find bats dead in the bottom of stored vessels. In one documented case, the calls of one trapped bat attracted others, escalating the negative impact of the situation. Bats also have been found dead after being impaled on barbwire fences or even caught in the strong velcro-like grip of burdocks.

Parasites

Bats host a rich diversity of parasites, including some that live on the outside (ectoparasites) and those that live inside the bats (endoparasites).[101] Ectoparasites include fleas, mites, ticks, bat flies (streblids or nycteribiids) and bedbugs. These are mainly small, some have wings, some do not. All of them mainly consume bat blood. The Naked Bat, a free-tailed species from Southeast Asia, hosts a specialized earwig (*Arixenia esau*). This insect is larger than the parasites and does not ingest bat blood but is known as a commensal, living on its host while eating some secretions and shed skin. To put the earwig in perspective, a creature of this size living on an adult human would be similar in length to the diameter of a dinner plate.

A close-up view of a winged streblid. These ectoparasites stand out on the bat's wings, but others could lurk unnoticed in the fur.

This flying Jamaican Fruit Bat has two orange streblid flies on its right wing and one on its left wing near its wrist.

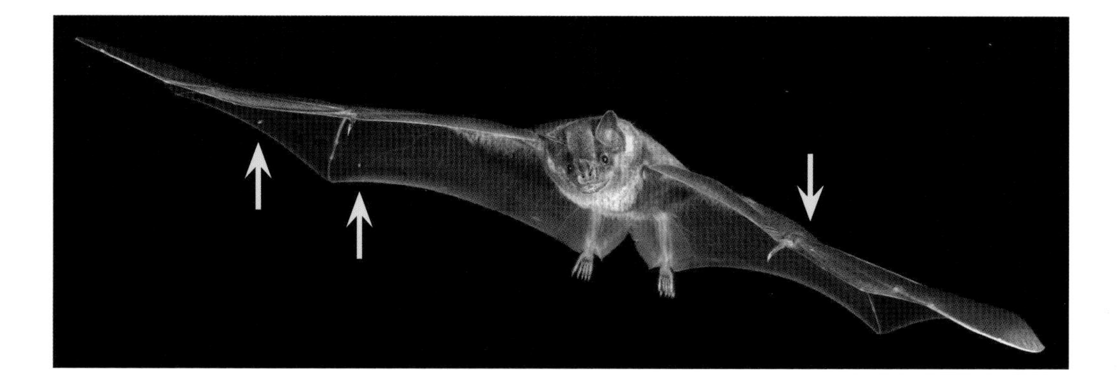

This Big Brown Bat died when entangled on a burdock (opposite). Young animals may be more vulnerable to such accidents than adults.

Ectoparasites can lurk in the fur, such as the nycteribiid bat fly on a Natal Long-fingered Bat.

A bedbug (dorsal and ventral views) commonly found on bats in Canada and the United States. These bedbugs may occasionally abandon bats and switch to humans as hosts.

Ectoparasites are usually insects or mites and may only climb on a host to feed. But some bat flies (streblids) shed their wings and enter a tear duct, or a hair follicle, while others embed themselves under a bat's skin. Bats living in groups in busy, dark and sheltered roosts are 'ideal' hosts for ectoparasites; those that roost in foliage and in the open tend to have fewer ectoparasites.

The 100 or so species of bedbugs are blood-feeding ectoparasites mainly of bats and birds. Some species are very host-specific, while others eat the blood of a range of hosts. A female Common Bedbug weighs about 5 mg and needs a 7.6 ng (nanogram) blood meal before she can reproduce. A 10 g bat roosting alone might have 1 g of blood, and a cluster of ten of these bats would provide more than enough blood for several bedbugs. Hot bats, those just returned from foraging, bleed readily and are good targets for bedbugs. But within 20 min of returning from foraging, the bats cool down, their blood pressure drops and a cool bat bleeds much less readily than a hot one. This can pose a problem for bedbugs that rely on bats for food.

Some bedbugs have switched from bats to humans.[102] A 50 kg (110 lb) human has about 5 kg of blood – a veritable gold mine for bedbugs. Furthermore, humans do not show the same temperature fluctuations as bats. Bedbugs living in the attic of a house inhabited by humans might just 'go downstairs' and sup on human blood. This switch in hosts could trigger an explosion of the population of bedbugs.

Polyctenid insects such as this (8 mm long) are close relatives of bedbugs and are normally ectoparasitic on free-tailed bats. They bear live young.

Common Tent-making Bats (above) occasionally harbour concentrations of larval mites that live in the follicles of sensory hairs on the neck. This follicle (right) is so packed with mites that it has become greatly enlarged.

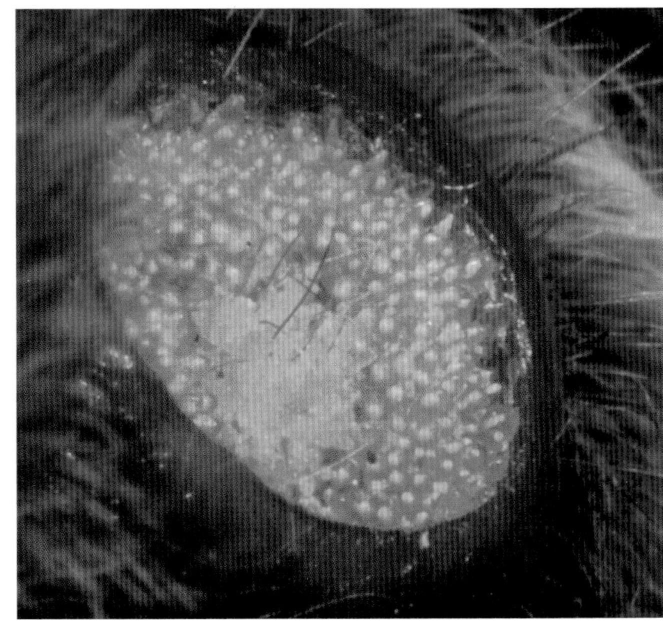

Small, orange mites (arrows) are found on the skin of different bats. On this Eastern Small-footed Bat there are mites on the ear and tragus.

Mites in a hair follicle of a Common Tent-making Bat make the point that a parasite normally living on the outside of its host may move partly inside. Other kinds of mites partially embed themselves in the skin of bats. Such mites occur on tropical species (Lesser Dog-like Bats – Emballonuridae) and temperate species (Eastern Small-footed Bats – Vespertilionidae).

We know much less about bat endoparasites, such as blood parasites or tapeworms. In yet another turn, we recently learned that host–parasite interactions involve not just host and parasite, but also the respective gut flora (microbiota) of host and parasite.[103] In some cases, the microbiota actually attract certain parasites, while in other cases they help the host defend itself from attacks by parasites.

A Western Barbastelle, with parasitic mites on the ears and wing.

A Lesser Dog-like Bat (6.5 g), with mites in the ear margins.

Wind Turbines

This aerial view shows five wind turbines in farmland on the north shore of Lake Erie (in Ontario).

Flying bats are killed at wind turbines when they collide with moving rotor blades or are trapped in the low air pressure behind the blades. In the latter case, the lungs implode, destroying lungs and blood vessels.[104] Individual wind turbines can in some cases kill more than 100 bats annually; this amounts to many hundreds of thousands or millions of bats killed at wind farms around the world every year. Some evidence suggests that this mortality has already reduced populations of bats. Although rates of bat deaths vary dramatically in space and time, there are some obvious patterns.

Bat deaths at turbines are most frequent during late summer and autumn and often concentrated during nights with warm temperatures and low wind speeds. This coincides with the migration and mating season of bats. Furthermore, warm nights with low wind speeds often coincide with high levels of insect activity. Bat species most often killed at turbines are those that fly straight and fast in open air space, relying on high-intensity echolocation calls to detect insects. At northern latitudes, species that are long-distance migrants and tend to roost in trees can comprise 75–90% of the fatalities in some areas. Examples are species of noctules in Europe and Hoary Bats and Eastern Red Bats in the United States and Canada.[105]

Why do bats approach wind turbines? They appear to be more at risk at wind turbines than birds, suggesting that, unlike birds, bats that approach are actively searching for something. The list of possibilities includes the presence of swarming insects or other

Viewed from the ground, turbines are impressive – as illustrated by this wind-energy facility located on the north shore of Lake Superior (Ontario). We found one dead Little Brown Myotis (inset) under one of the turbines at this facility. The bat had suffered a compound fracture of its left humerus.

bats (and/or their scent marks), perhaps potential mates. Mating pairs of Eastern Red Bats have been found dead below wind turbines, supporting the mate search hypothesis. Some bats also feed around turbines, while others fly past without feeding or stopping.

JR and colleagues studied bats at turbines using a new entomological lidar, LUMBO (Lund University Mobile Biosphere Observatory).[106] The machine monitors flying objects as they cross a laser beam and can, in practice, detect a mosquito at 2 km or more away. Most importantly, it works independently of natural light. We used it at night to monitor insects and bats in the air near a wind turbine 100 m or more above the ground. We combined LUMBO with a bat detector that registered feeding buzzes and other calls, including songs. With this dual approach we could evaluate happenings at the wind turbine. Common Pipistrelles came to feed and to mate. Sometimes a male monopolized the turbine and spent most of the night singing, with little or no feeding.

When a wind turbine is a place where bats find food and mates, there seems little point in trying to convince them to go elsewhere. Mitigation of the impact of turbines on bats must depend upon other methods. The scheme most frequently used in Sweden is switching off the rotor when the risk is highest. In practice, this means between sunset and sunrise in August and September on nights when the air is warm and winds are light. This strategy reduces the death rates of bats at turbines without a forbiddingly high production loss. At wind farms in the United States and Canada, setting turbines to activate only when winds exceed 9 km per hour also reduces bat mortality. The challenge now is convincing decisionmakers and energy producers in countries outside Europe and North America to adopt similar schemes for the benefit of bats.

Fortunately, there can be a positive side to bat deaths at wind turbines. Researchers measured the incidence of rabies virus in carcasses of Silver-haired Bats and Hoary Bats retrieved from under turbines. They found that 1% of 96 Silver-haired Bats tested positive for rabies, but none of 121 Hoary Bats. The low quotient encountered in this important study helps to put the association of bats and rabies in perspective (see also page 210).

Estimating population sizes is a recurring challenge for bat biologists.[107] Knowing population size is fundamental in conservation plans, but for most species we lack robust data. In an innovative approach, researchers used genomic analysis to measure effective genetic population size in two samples of Hoary Bats collected from under turbines. The 2009–10 and 2017–18 samples each contained 93 bats. The analysis suggested a doubling of the effective genetic population in the last 100,000 years. They estimated the effective breeding population from the genetic data and reported an increase between the first and second samples. This could be welcome news.

In the meantime, on a worldwide scale, wind turbines kill more bats than any other single known threat.

Light pollution

There is a worldwide explosion of artificial lights, particularly since the introduction of LED (light-emitting diodes) lighting, which give more light and consume less electricity.[108] Lighting at night (LAN) has turned night into day in many locations. Disruption of the natural light–dark cycle negatively affects life, from single cells to whole ecosystems. Light in the wrong place or at the wrong time is light pollution, spread by humans and harmful to the environment. For nocturnal animals such as bats, darkness

A Brown Long-eared Bat roosting on a bell in a church steeple. *JR*

offers protection against predators. Light pollution can directly affect their survival and reproduction. In parts of the United States, Little Brown Myotis are less active in lighted than unlighted situations, while Eastern Red Bats and Hoary Bats often forage at lights. Big Brown Bats and Silver-haired Bats are less affected by LAN. For some species LAN seriously degrades otherwise suitable foraging habitat.

Nearly all bats hide by day, are active at night and avoid lights whenever possible. However, the details of their responses depend on the species, the bat's predators and the situation. Fast-flying, agile species may handle predators better than slower, less agile ones. Faster-flying bats often emerge earlier when there is still some light, and an abundance of insects. This carries with it some risk of being killed by a bird (see page 172). Postponing emergence until full darkness is safer, but then there may be fewer insects – another example of a trade-off. The fashion to illuminate old cultural buildings such as churches and monuments from the outside at night has been important for some bats. The lighted buildings had often been preferred bat roosts prior to such developments. Brown Long-eared Bats are an obvious example. Floodlights on buildings can seriously compromise the safety of the bats roosting inside, exposing them to predation by waiting owls and hawks. This problem has severely affected some populations of bats. At night, artificial light affects bats in many other ways. Rows of lights act as barriers to small horseshoe bats and long-eared bats, restricting their movements both locally and on a larger scale. Lights on ponds and at other sites where bats take water can keep them from drinking there. Most importantly, bats of nearly all species avoid lit areas for nursery colonies. Young bats practising flight and echolocation are extremely vulnerable to predators. Here there is no room for compromising safety, and females move to the darkest possible roosts.

Some species often hunt in concentrations of insects at streetlights (see page 76). Here bats usually patrol the lamps from above or the side, entering the light cone only during quick attacks on prey. In such cases, the lighting can be good for the bats, increasing the rates at which they encounter prey and interfering with moths' defensive behaviour.

Rows of streetlights that are a barrier to movements of sensitive species of bats can lead to fragmentation of habitat and, eventually, local disappearance of these species. Extensively lit areas typically have a few opportunistic bat species or none at all: light pollution has important consequences for the distribution and abundance of most bat species.

How can we mitigate the use of outdoor lighting and protect bats? Bats' eyes are less sensitive to red than to white light, so using red lights might be better for bats. Using timers to minimize the amount of time outdoor night lights are on also could benefit these animals. If lights are turned on by motion sensors, they might have less impact on bats. But it seems that in some situations, darkness is the only solution.

Two Swedish churches used as roosts by Brown Long-eared Bats. A church in Ulricehamn was used by the bats until lights were installed in 2016, when the animals left. Note the bat flying outside. A church in Falköping (inset) without lights but with a colony of this species. *JR*

A Little Brown Myotis showing classic symptoms of white-nose syndrome

A world without bats?

Before March 2006, the very idea that we could lose millions of bats to the effects of an introduced fungus was beyond our comprehension.[109] But in March 2006 bat biologists from New York State were making their annual counts of bats hibernating underground. At three caves, what they had expected was the challenge of counting the numbers of Little Brown Myotis hanging in clusters from the ceilings of the sites. What they found instead was piles of dead bats on the floor, beneath the ceilings the bats had used for so long. Some of the surviving bats had what looked like cotton fluff on their muzzles, the hallmark of white-nose syndrome (WNS).

To the utter dismay of bat biologists everywhere, over the next five years WNS progressed steadily outward from the original sites, and we lost more and more bats.[110] By 2010 two favourite sites in abandoned mines in southern Ontario that each had harboured over 30,000 hibernating Little Brown Myotis now were almost entirely without bats. In Nova Scotia, summer and winter sites that once had large colonies of this species now had none.

The culprit was the European strain of a cold-loving fungus, common there in caves and mines. Somehow the fungus got into a few sites in the United States. One key point lies in the phrase 'cold-loving', which makes underground bat hibernacula ideal. A second key point is that at night, bats swarm at these sites from early August through autumn. Swarming means that tens or hundreds of bats visit, including adults and subadults. Swarming may show young where to hibernate. Swarming also is mating time, so there is considerable bat-to-bat contact, increasing exposure to the fungus.

The fungus interrupts the cycle of hibernation. Instead of arousing from torpor once every few weeks, bats arouse several times a week and exhaust their body fat well before the end of winter. This dooms infected bats. Transmission of the fungus is bat-to-bat.

The rapid spread of WNS made it starkly obvious that Little Brown Myotis moved among hibernation sites far more often than we had suspected. Furthermore, these bats covered longer distances than we had documented by band recoveries. Not all species of bats appear to have been affected by WNS; in some cases (Eastern Red Bats, Hoary Bats, Silver-haired Bats) they probably escaped because they do not use underground sites. WNS does not appear to survive in summer nursery sites which are too warm for the fungus.

By 2010, some wildlife biologists in the United States expected that by 2020 Little Brown Myotis would be extirpated from northeastern North America. In 2020 the species persists in much of its former range but in very reduced numbers. In February 2019, during a visit to an abandoned mine in southern Ontario, biologists counted 4,500 Little Brown Myotis. This was uplifting, but a far cry from the 30,000 that had been there before WNS. In 2020, researchers showed that supplementary feeding of Little Brown Myotis was a way to promote recovery from WNS. The work identified ways in which we could increase the chances of individual Little Brown Myotis surviving.

European bats are not so affected by the local strain that causes WNS, but they presumably have had a longer period to develop protection. The slow rate of reproduction means that populations of Little Brown Myotis will be slow to rebound, but there is some evidence of recovery at some of the hibernation sites.

Humans have current experience with a pandemic, but our experience pales in comparison to the impact of WNS on bats.

Global change

Does climate change pose a threat to bats?[111] At northern latitudes, ongoing climate change will mean higher annual temperatures, shorter, milder winters and longer summers. Rainfall and weather patterns will also be affected. In Europe bats may have shorter hibernation periods but in warmer hibernacula. Reproduction may be facilitated by warmer and longer summers but will also depend on other factors such as rainfall. More species of bats may occur regularly in parts of Europe where they currently are rare. In recent work with Bechstein's Bat in Germany, warmer summer temperatures resulted in larger body sizes and increased risk of mortality. Here the data come from long-term studies of individuals marked with PIT tags.

Now, our challenge is to compare the predictions with observations from the real world. In Scandinavia, as in the rest of Europe, there has been a dramatic spread of pipistrelle species.[112] Thirty years ago, only the Soprano Pipistrelle (see page 125) was common in Scandinavia, but now there are three: both the Common Pipistrelle and Nathusius' Pipistrelle have become well established too. Furthermore, a fourth species (Kuhl's Pipistrelle) seems to be on its way, currently spreading fast northwards in Central Europe. For pipistrelles, increasing abundance and extended range are likely to be direct effects of climate change, but this has not yet been explored.

We also have seen a dramatic increase in the number of bats in many hibernacula across Europe, including Scandinavia.[113] This seems independent of pipistrelles, and mostly reflects mouse-eared bats (*Myotis* spp.) but also long-eared bats and barbastelles. The increase reflects bats that normally hibernate underground in caves and mines. But here there are few indications of population increases beyond the hibernation counts. In pipistrelle species, changes in abundance also occur in summer. There is evidence of a dramatic increase in the number of hibernating barbastelles at hibernation sites in Poland. Here the bats appear to come from another nearby, slightly warmer site. Change in numbers of hibernating Western Barbastelles appears to reflect a move to cooler hibernation sites. It is possible that the increasing numbers of mouse-eared bats at some hibernation sites in Europe also reflects movements to remaining cool hibernacula.

In southern Sweden, populations of Northern Bats – the northernmost-occurring of all bat species – have declined by more than half, while pipistrelles have expanded dramatically.[114] But the decline in Northern Bats could reflect changes in street lighting. Thirty years ago Northern Bats were commonly observed feeding on insects attracted to mercury-vapour streetlights. Changes in streetlighting made the lights less attractive to insects, and activity of Northern Bats declined by 60%, reflecting lower food availability.

Quantification of emergences by Brazilian Free-tailed Bats at five maternity cave colonies over 11 years indicated that bats emerged significantly earlier during drought compared to moist years. These researchers used data from the NEXRAD network of Doppler weather radars. Earlier emergences could provide more access to insects, albeit with more risk of predation (see page 172).

In Colorado, researchers studied bats coming to drink at a 1.5-m-diameter artificial water source 600 m from a nursery roost of Fringed Myotis.[115] They captured and PIT-tagged 24 adult females and deployed a plate antenna in the artificial pool. The plate antenna detected drinking events by PIT-tagged individuals. Lactating female fringed

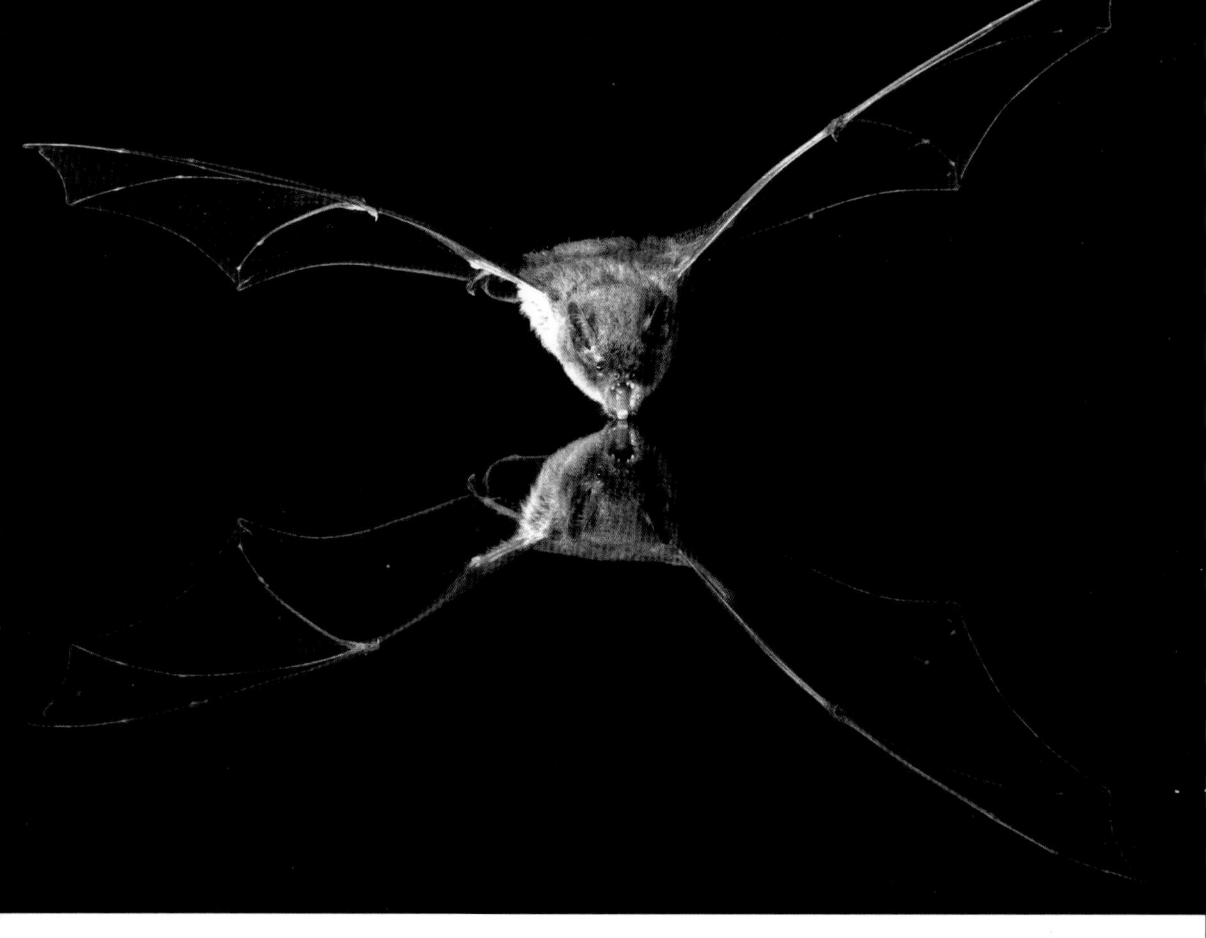

A Zulu Serotine drinking at a tank in South Africa. By placing an antenna under the water, researchers could monitor drinking by PIT-tagged bats. This opens new avenues for documenting how and when bats use space.

myotis made 13 times more drinking visits than non-lactating females. These findings suggest that a reduction in surface water (drinking sites) could negatively affect some bats.

Using mitochondrial DNA, researchers in South Africa compared changes in the genetic structure and population stability in six species of bats in the last 20,000 years.[116] They chose three species associated with forests (Cape Hairy Bat, Botswana Long-eared Bat, Lesser Long-fingered Bat) and three species that are habitat generalists (Swinny's Horseshoe Bat, Banana Pipistrelle, Dusky Pipistrelle). The data they considered included natal range, and dispersal ability. All six species showed evidence of population expansion, suggesting no effect of climate change. None of the species demonstrated a decline in population since the last glacial maximum.

How will global climate change affect bats? It depends… on the species, on the location, on what they eat. Arguably, changes may be more associated with habitat loss and disturbance.

Keeping bats away

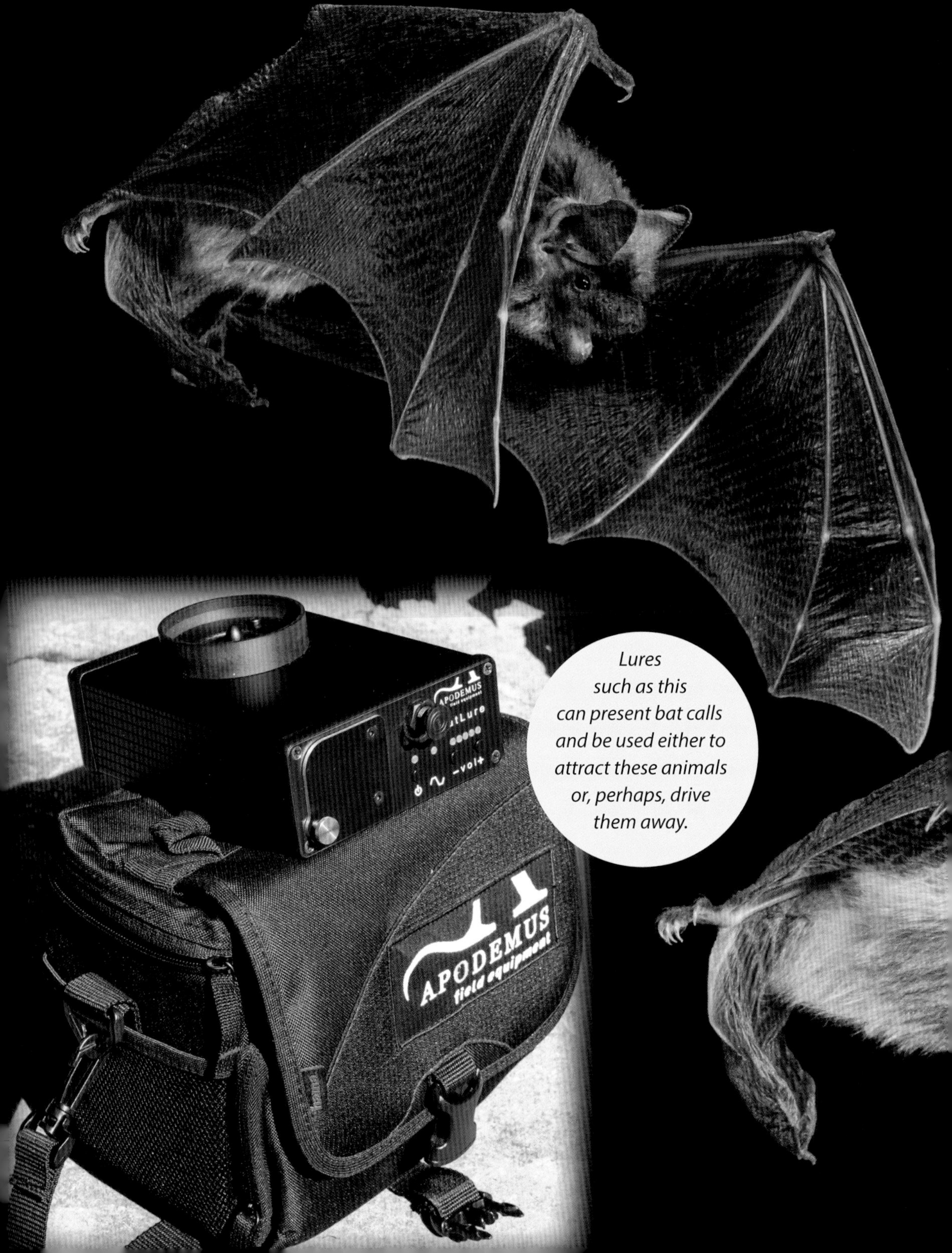

Lures such as this can present bat calls and be used either to attract these animals or, perhaps, drive them away.

Biologists had the impression that bats avoided areas with ambient ultrasonic sound. Moreover, some appeared to modify their echolocation calls to minimize interference. Researchers out of Bristol (UK) explored the possibility of using acoustic deterrents to minimize bats' exposure to dangerous situations.[117] In areas where they had encountered high levels of bat activity, the researchers broadcast white noise (20–100 kHz) in a 15–30 m space. The broadcasts overlapped the bandwidths of the echolocation calls of Common Pipistrelles and Soprano Pipistrelles, as well as other bats in the study area (mouse-eared bats (*Myotis*), noctules (*Nyctalus*) and house bats (*Eptesicus*)).

The research revealed three basic responses to the acoustic deterrent treatment. First, bats flew faster through the treated areas and used more tortuous flight paths. There was less overall bat traffic (measured by echolocation calls) where the acoustic treatment was broadcast. Furthermore, Soprano Pipistrelles produced fewer feeding buzzes and social calls, while Common Pipistrelles showed no obvious change in behaviour in the deterrent area. Soprano Pipistrelles also changed their echolocation calls, both start frequencies and bandwidth, when flying in the deterrent area.

In short, acoustic deterrents significantly affected bat behaviour, from what they said (echolocation calls), to their flight speeds and flight paths. Overall bat traffic was reduced in areas of deterrent presentation. Not all species responded the same way, but general bat aversion to ultrasonic deterrents implies that there is significant potential for reducing bat exposure to turbines or areas of high vehicular traffic.

As usual, high-frequency sounds (ultrasonic) do not travel as far as low-frequency sounds, perhaps limiting areas over which deterrents can be effective. This might be overcome by using more speakers deployed over a larger area.

Big Brown Bat in dynamic flight.

12 Bats and people

Emergences of bats from roosts can be spectacular and attract people's attention. In Austin, Texas, Brazilian Free-tailed Bats that emerge from their roosts in the Congress Avenue Bridge are an important tourist attraction.
Photo by Amanda Stronza, Bat Conservation International

A WuFu coin (36 mm diameter) from Canada shows five bats surrounding the Chinese symbol shou (longevity). Each bat represents a blessing – namely longevity, wealth, health and composure, love of virtue, and a peaceful death in old age.

Many people are curious about bats, large colonies of which can be a tourist attraction. Opportunities to observe live bats often draw a lot of attention. There are many public exhibits about these animals, which also can have a strong positive influence on the public view of bats. At the Royal Ontario Museum in Toronto (Canada), since it opened in 1988, The Bat Cave exhibit has been a top choice of visitors. At the main airport in Israel, an exhibit of Jens' photographs of bats attracted a great deal of attention. Bats are positive symbols in many cultures and there are myths and stories about bats all around the world. Bats do some wonderful things: they can see in the dark and they can fly, powers that have long caused them to be associated with witchcraft.

Bats are symbols of good luck and fortune in Taoism. To our knowledge, negative attitudes about bats do not appear in most major religions including Islam, Hinduism, Buddhism and Judaism. To the Taironian people who lived in what is now northern Columbia and Venezuela, vampire bats symbolized the fertility of women. Has she been bitten by the bat? Is she of child-bearing age? In Papua New Guinea, the connection between bats and the fertility of women also is clear.

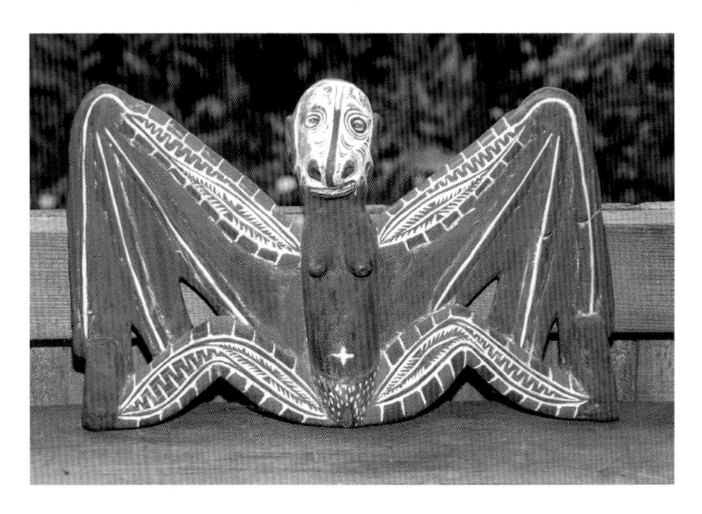

A bat symbol from Papua New Guinea.

Attitudes towards bats

Some people fear bats. Why? Unlike some other mammals, bats are essentially harmless to humans. Their elusive behaviour and strange appearance have always fascinated us. Fear of bats is a learned tradition in European/American culture that originated long ago in the church. The fear was later reinforced through beliefs in witchcraft and other cultural expressions, such as poetry, fiction, films and the modern Halloween tradition.[118]

A devil with bat-like wings. *JR*

Fear of bats has been exported to the rest of the world, including many countries where such ideas were previously unknown.[119] In the Bible, the bat is portrayed as an unclean animal because of its ambiguous appearance. It is neither bird nor beast and cannot be eaten or even touched. Bats were chosen as icons of the Devil during the Middle Ages, partly because dark wings contrasted well with the white wings of angels. This placed bats as the undisputed enemy of God in Christian philosophy and iconography. Accordingly, the Devil was equipped with bat wings in the thirteenth century, first in religious artwork, then emphasized in influential writing such as Dante's *Comedia* ('The Divine Comedy'). In this book, Lucifer (Satan) complete with bat wings, resides at the bottom of Hell. Church paintings showing devils with attributes of the bat appeared and became common in southern Europe in the thirteenth century, and in Scandinavia about 100 years later.

順

A bat symbol is prominent in a Taoist shrine in Taiwan.
JR

Bat-winged devils disappeared from churches during the Romantic period (early nineteenth century), a time when they became popular in other, mostly non-religious, artwork. In architecture, bats were included mostly because of their magic properties. They could protect buildings and their inhabitants from witchcraft and bad luck in general.

Above: Ceiling painting in a Swedish church, note the bat-like wings. Below: In Sweden, the bat symbolized protection of the building. *JR*

According to Scandinavian folklore, magic properties historically gave the bat an important role. It was used to facilitate everyday life in many ways, from love affairs, to hunting and fishing, gambling, business, protection and healing. Despite the message from the Church, the bat was respected and often used by people to adorn things. However, in folklore and witchcraft, bats were used in secrecy. This is because bat symbolism was strictly forbidden because of the clash with Church doctrine. The historical importance of this has probably been seriously underestimated. The Church's fight against witchcraft, including the Devil and the Bat, eventually resulted in the wider witch-hunt, one of the worst persecutions in European history. Thousands of women and some men and children were killed and burned in the course of this potent moral panic.

The real importance of bats in Scandinavia is illustrated by a silver engraving on a pistol of the Swedish King Gustav Adolf, signed 1603. Although he was a powerful leader and the undisputed head of both state and the Protestant Church, he evidently relied on God but also on the bat. The bat officially may have symbolized the Devil, it also had another secret, parallel life which may have been more important. Despite the bat, however, Gustav Adolf was killed in a battle during the Thirty Years' War.

King Gustav Adolf's pistols with close-up showing a bat outline under the trigger guard.
JR

In several hundred museum records about the use of bats in Swedish folklore, we did not find a single claim that bats were dangerous, aggressive, poisonous or harbour disease! The fear-of-bats attitude did not prevail in the countryside before the Middle Ages. The same may have been true elsewhere, as indicated by carvings of bats in the cathedral of Notre Dame in Paris.

A carved bat inside Notre Dame Cathedral, Paris. *JR*

Some large colonies of bats roost in artificial structures where they can be unwelcome because of the noise they make, and/or their deposits of faeces and urine. Human attitudes towards house bats vary widely. When bats are associated with diseases that affect people, our response usually is negative. In India, bats commonly roost in temples and monuments, sometimes with protection associated with people's religious beliefs. Peaceful coexistence can prevail, even when many bats are involved. Spreading negative attitudes about bats appears to be a Christian specialty. Today people in India recognize that European and American tourists' fear of bats hampers the tourist business. To maintain tourism, this has led to destruction of bats at some archaeological sites in India.

Above: Detail of a misericord at a church in Hereford, England.
Below: Bats shown on a tomb in Père Lachaise Cemetery, Paris. *JR*

A Greater Mouse-tailed Bat flying in the Feroz Shah Kotla mosque in India is one of a large number of this species that roost there. *JR*

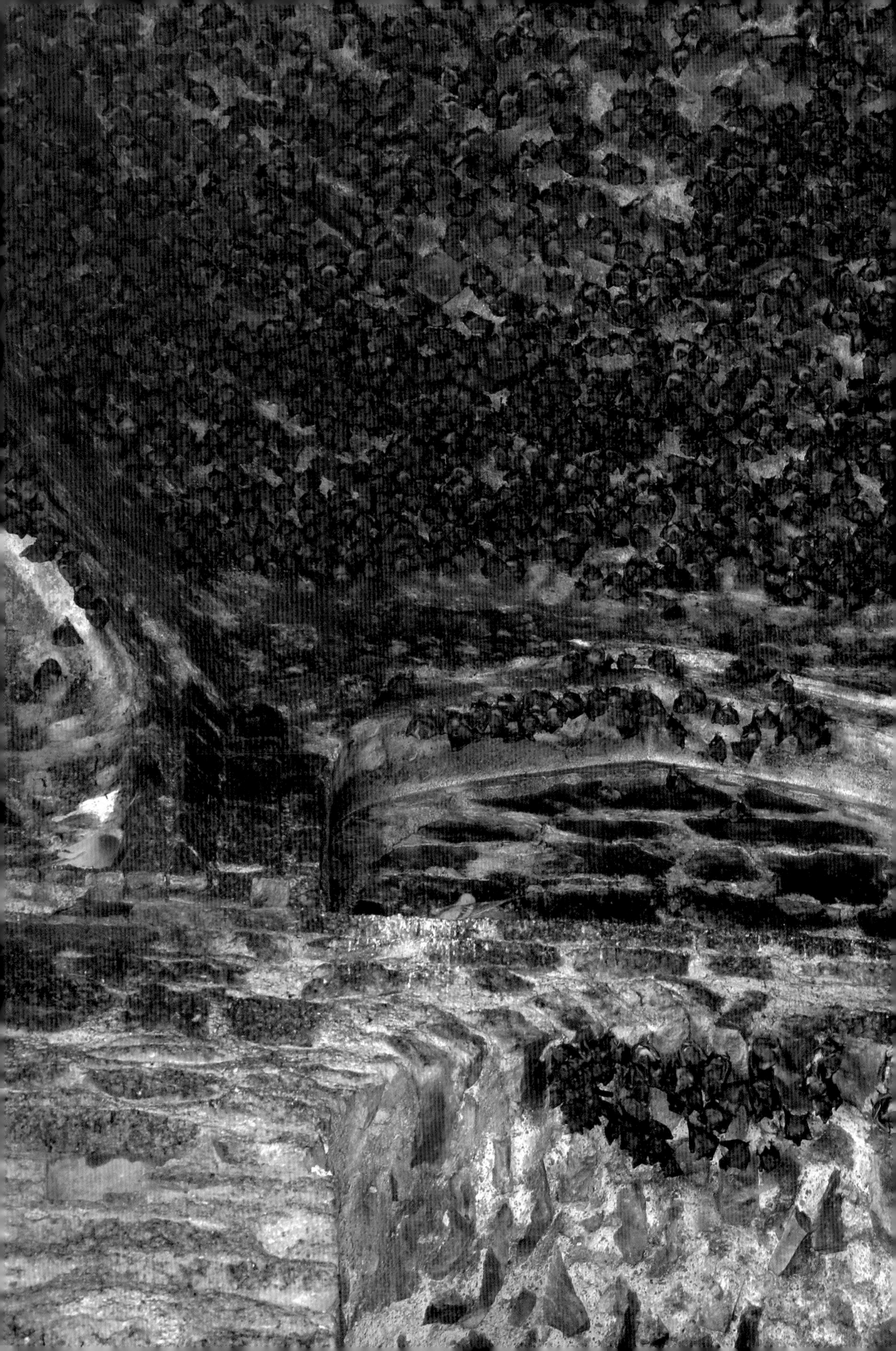

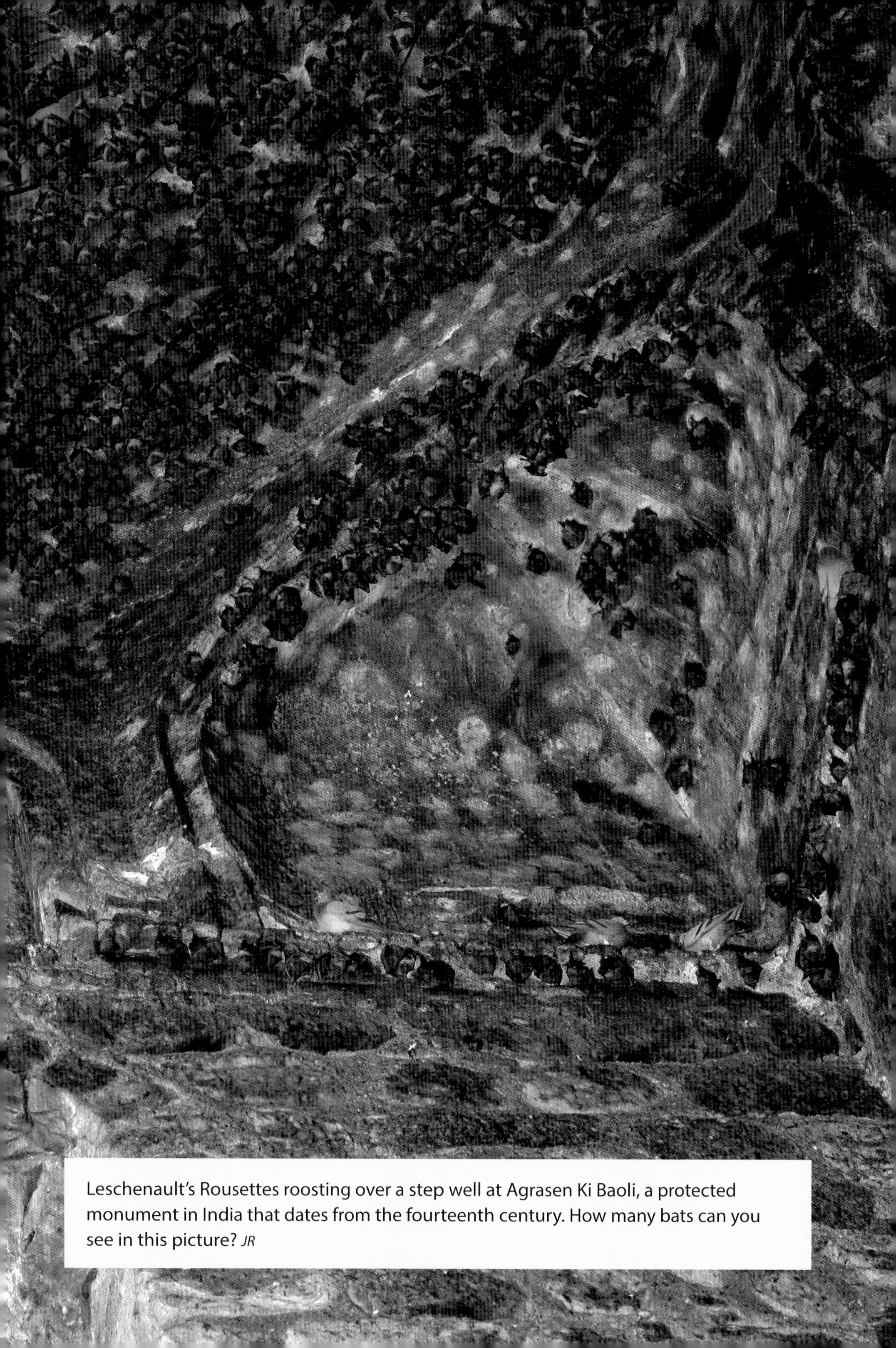

Leschenault's Rousettes roosting over a step well at Agrasen Ki Baoli, a protected monument in India that dates from the fourteenth century. How many bats can you see in this picture? *JR*

Bats and disease

Widespread belief that SARS-CoV-2 virus (which causes COVID-19) is associated with bats has exacerbated negative public attitudes. Media in Europe and North America have broadcast this view around the world, even though it is inappropriate and inaccurate. The appearance of SARS-CoV-2 virus was linked to bats on the grounds that it is closely related to SARS-CoV, which caused SARS. COVID-19 resulted in a pandemic that had far-reaching impacts on humans. Just when, or how, SARS-CoV-2 spread from bats to humans remains unknown and open to speculation. Many zoonotic diseases have jumped (spilled over) from wild animals to people over the course of history, sometimes with disastrous consequences. The long list includes AIDS, Ebola, SARS, MERS and more. These spillover events reflect expanding human populations and encroachment on natural habitats. As shown by COVID-19, zoonotic diseases can spread rapidly through the human population, with disastrous effects. But bats themselves do not cause spillover events, or COVID-19.[120]

If bats were important sources of zoonotic diseases for humans, biologists who study them should be valuable 'canaries in the coal mine'. To assess this situation, in the wake of the SARS outbreak a group of researchers analysed the blood of 90 bat biologists registered at the 2005 meeting of NASBR, the North American Symposium on Bat Research.[121] Analysis of the 10 ml samples revealed no evidence of antibodies against SARS-CoV N protein in 89 of 90 samples. Further analysis showed that the single positive sample reflected a CoV unlike SARS. The biologists who volunteered blood had worked at sites around the world, including China, and studied a wide range of bat species. We still do not know how SARS-CoV-2 moved from bats to humans.

Rabies used to be the most feared disease associated with bats.[122] Rabies is caused by another virus, although not one closely related to SARS-CoV viruses. Rabies affects the nervous system and, once an animal shows clinical symptoms, the disease usually is terminal. Worldwide, about 59,000 people a year die from rabies but in much of the world the impact is much less. For example, in Canada between 1970 and 2019 there were nine human deaths from rabies, seven caused by strains associated with bats. The other two cases involved strains of rabies associated with dogs. In Africa and Asia, dogs

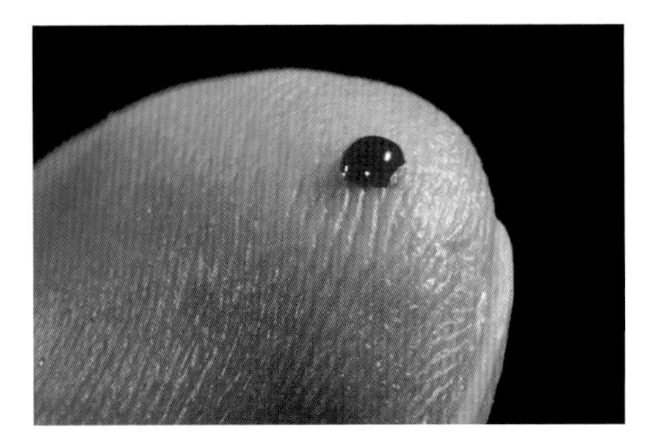

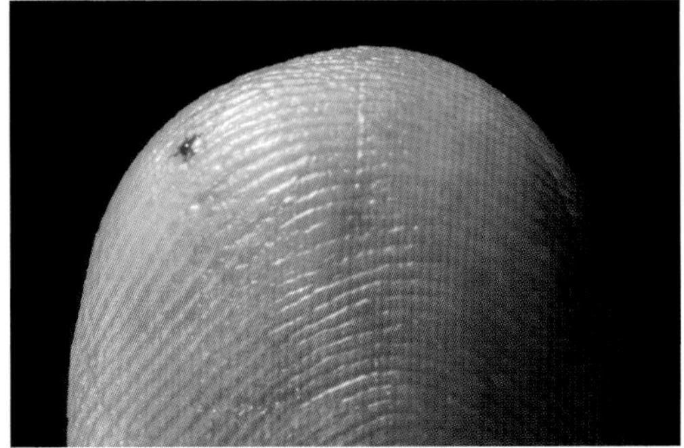

Drop of blood at the site of a puncture wound caused by a defensive bite of a Big Brown Bat. Five minutes later there was little evidence of the wound.

are the main source of rabies in humans. Outside of the Americas, spillover of rabies (and related viruses) to people is uncommon, but the risk remains. Post-exposure rabies vaccinations provide effective protection from the virus. Vaccinations can also protect people who work with bats and other wild animals from rabies. Annual tests of level of protection (titer) indicate when it's time for a booster shot to maintain protection.

Rabies is usually transmitted by a bite. Animals suffering from rabies typically have virus in their saliva and paralysis of their throat muscles may limit their ability to swallow the saliva they produce. These animals can appear to foam at the mouth, and a bite can transfer virus into the wound. But how do you know you were bitten? When the bite is delivered by a dog or a cat, there usually is an obvious wound. Bats are much smaller than cats and dogs, and their bites readily pass unnoticed. On 10 March 2021, MBF was bitten by a Big Brown Bat (14 g) that was annoyed with him. The puncture made by the bat's upper canine tooth was minute. Within 5 minutes, the bite mark was not obvious. MBF has had his pre-exposure shots and has good protection, but he should have worn a glove. This unsolicited bite was hardly the bat's fault.

Here are two take-home messages. First, do not handle wild animals, including bats. Second, if your job requires you to handle wild animals, get pre-exposure vaccinations and make sure your protection is up to date, and always wear gloves! Do not expect to find a bite mark.

Bats are also associated with other viruses.[123] Hendra and Nipah viruses are shed in bodily fluids of some pteropodid bats and have been transmitted directly to people. They also have been transmitted indirectly to people via horses (Hendra) or pigs (Nipah). The five species of Ebola virus cause haemorrhagic fever that can be lethal.

Too often, the connections of bats to many viruses are unclear. There is no indication of a connection to bats in, for example, the spread of Marburgvirus, Lloviu virus and influenza viruses. Lloviu occurs in some Egyptian Fruit Bats and in Schreiber's Long-fingered Bat, and influenza viruses (*Orthomyxoviruses*) in other species of bats. Monitoring bats for viruses is a work in progress. In many cases we lack details necessary to understand the nature of the risk(s) the viruses might pose, or how they might get from bats to humans. Overall, bats are no more prone to harbour viruses than birds and other mammals.[124]

Histoplasmosis (histo), a fungal disease of the lungs, is also associated with bats. Spores of the fungus often occur in bat droppings, as well as the droppings of pigeons and chickens. Although the symptoms of histoplasmosis are usually mild and flu-like, it can be much more serious. This disease is common in many parts of the world, and people often are exposed when they explore natural or artificial bat roosts. A mask that filters out particles larger than 2 μ in diameter can protect the wearer from exposure to the spores of the fungus that causes the disease.

As a group, bats have extraordinary abilities to neutralize dangerous viruses, but bats have no higher rates of infection by viruses than other mammals and birds. Blood-feeding vampire bats are the only bats that bite humans to feed. However, all 1,400+ species of bats have teeth and will bite in self-defence.

We can learn a great deal from bats, but for most people they are best appreciated at arm's length. We hope that current fear of COVID-19 does not spillover into and strengthen more traditional public fears of and concerns about bats.

Bats as symbols

In one cemetery in Paris (France) bats appear on some of the tombs perhaps reflecting a general association of bats with the afterlife and the underworld. The representations may be silhouettes or more detailed. As usual, wings identify the representations as bats.

Bats shown on two tombs in Père Lachaise Cemetery in Paris. *JR*

Bats often appear on postage stamps. In 1978, Germany issued a postage stamp bearing the image of a fossil Eocene bat (*Palaeochiropteryx tupaiodon*). The bat was one of many species known from the famous Messel Pit – a disused quarry rich in fossil remains – providing important glimpses to the fauna of the time. The bat stamp was one of a series that highlighted this important fossil deposit.

Three stamps showing bats, two associated with the city of Valencia in Spain (despite one being a German stamp). Note the differences in the style of the bat silhouettes between the Spanish and Colombian stamps. The stamp on the right depicts the fossilised remains of a Small Messel Bat.

Other stamps that feature bats tell different stories. Bats on the coat of arms of Valencia, a city in Spain, recognize a major event in Spanish history. The Moors had taken Valencia from the Goths in 711 CE and held it until they were defeated by James I of Aragon in 1238 CE. On the eve of the battle for Valencia, a bat flew into the king's tent. This was taken as a good omen and so a bat then appeared on the city's coat of arms. The bat symbol also appears on the logos of the city's soccer teams. The six silver bats on the stamp from Colombia are from Santa Fe de Antioquia, a city in northwestern Colombia. Some will see a resemblance between the Valencia bat silhouettes and the Bacardi bat. In other stamps the lines are less clear. The bat from Moldova is accurately labelled, but the one from Wallis and Futuna is not.

Two stamps depicting bats. The example from Moldova shows a species that occurs there (see page 141). The 'bat' from Wallis and Futuna has a fox's head attached to a generic bat body. This clearly is not the Peka, a local species of flying fox.

Bats also appear on coins. The American 25 cent piece shows a flying fox from American Samoa. The Polish 20 zotl piece shows a Lesser Horseshoe Bat. On both coins, the representations of the bats are reasonable, although without the label, identification could be a challenge.

Two bat coins shown heads and tails. The American 25 cent (quarter) coin has a Samoan Fruit Bat on the tails side. On the Polish coin, a Lesser Horseshoe Bat is tails.

Bat on a Napolese oil lamp.

There is a bat on an oil lamp displayed in the Naples National Archaeological Museum in Naples. The piece came from Pompei. While wings identify the animal as a bat, clearly a male, the significance of the association between the bat and the lamp is not clear.

A Chinese snuff bottle with bats. The bats are shown in red (the colour of joy), and carry coins.

Bats appear on the insignia of many different military units around the world. A bat is the official insignia of the United States Navy Torpedo Squadron 27. In World War II this unit was based in the South Pacific. Chosen to reflect nocturnal flying, the flying fox on the insignia is loosely based on an epauletted fruit bat. The bat was chosen for its wile and cunning and, as a last resort, as food for a lost pilot. A bat also adorns the badge of the Royal Air Force No. 9 Squadron, a British bombing unit from World War II. The badge had been approved by King Edward VIII in November 1936. That squadron's motto was *per noctum volamus* (through the night we fly).

The official insignia for the US Navy Torpedo Squadron 27.

13 Bats as beings

In recent years, new approaches to studying bats have enriched our knowledge of these animals and revealed previously unknown details about their lives. Two examples stand out. First, as noted on page 114, Common Vampire Bats in a social unit (colony) groom one another and even share blood meals by regurgitation. Proximity tags revealed long-term associations among individuals ('friends') in their social units. The researchers identified 'friends' by assessing inter-individual contacts among group members. From band recoveries, biologists had long recognized that some bats repeatedly shared roosts. But the ability to recognize and track known individuals was key to the new findings because it opened a window on what bats might do. The data demonstrated complexities of social groupings in bats that formed 'fission–fusion' societies. Individuals in groups may often change roosts and roost-mates = fission–fusion. The data also reveal that interactions extend beyond a particular roost to networks of roosts. This information is further enriched by knowing that individual bats can learn from observing other group members (see page 158).

Second, is the demonstration that at least Egyptian Fruit Bats use cognitive maps of their surroundings to locate roosts and food. What's particularly important here is the scale over which the bats operate. Some tagged individuals had home ranges of 60 km^2. Furthermore, young animals quickly learned their home ranges. Against this backdrop is evidence from other species that while some individuals are relatively sedentary, perhaps travelling 12 km among sites, others cover considerably larger distances. Banding studies from the 1950s through to the 1970s showed that band recoveries were most often in the roosts where the bats had originally been banded. PIT tags and Motus tags extended our perspective, revealing that even species weighing just 8 g cover tens to hundreds of kilometres between summer and winter roosts. Learning that some Lesser Long-nosed Bats covered at least 200 km in a night's feeding was astonishing.

When information about bats' social lives and use of space is combined with data on their longevity, we can learn and surmise a great deal about the complexities of their lives. It is exciting to catch banded Little Brown Myotis and, from band numbers, see that your sample of ten bats includes some that are 5 years old, others 10, 20 and even 30 years old in the wild. Exciting changes to ecstatic when you find that one bat also has a Motus tag and has travelled over 100 km between Motus stations in one night.

It can be all too easy to underestimate other individuals. We know this when the 'others' are colleagues or pets, yet we often fall into the trap. When you are sure that a bat you have just banded could not possibly cover tens of kilometres in a single night, then you are unlikely to find that they do. At one level, if you do not have the means to document how they spend their days and nights, you may not give it a second thought. In the 1960s we probably wrote off some outlying records of the movements of banded individuals as misread band numbers.

Donald Griffin had a strong interest in the question of animal awareness and published a book on the subject. The book was not well received by some colleagues, who did not believe that other animals were 'aware' as humans are. Don pointed out that if you

Van Gelder's Bat (18 g) occurs in parts of Central America, eating mainly insects and roosting in hollows. Its closest relatives are Pallid Bats (see pages 158–9).

did not expect other animals to be aware, you would not find evidence that they were. Likewise, when reads of PIT tags reveal that a bat has travelled tens of kilometres from one roost to another (see page 140), you may suspect the record. The same is true of Motus tags – and either system can generate errors. But when you realize that most of the data are accurate, you must look at your study animals in a new light.

The depths of a COVID-19 winter provided an interesting opportunity to get closer to bats. In spite of many years of studying these animals, it had been easy for us to focus on immediate research questions and not pay as much attention to the animals as we might have. This also meant overlooking lots of advice to the contrary. In her 1968 book *Ethology of Mammals*, Professor R.F. (Griff) Ewer noted that the best way to learn about the behaviour of mammals was to live with one. This is well known to anyone who has cats and dogs as pets (other animals as well). In his aforementioned 1978 book *The Question of Animal Awareness*, Donald Griffin explored various lines of evidence that humans were not the only animals to be aware of their surroundings including places, objects, other animals and other people.

Bat researchers who have worked with captive trained animals are quite aware of individual differences in species. The significance of inter-individual variations in behaviour expand our knowledge about the complexity of bats. The recent developments in research techniques have revealed that bats are more astonishing than expected. This is true even before you consider their capacity to neutralize viruses and avoid the negative impacts associated with them.

We close with some brief comments about two species in particular. In a sense, both Common Sheath-tailed Bats and Percival's Trident Bat epitomize the state of our knowledge. We know that each appears to represent just one species. Both eat insects, with the Common Sheath-tailed Bat perhaps eating more beetles than Percival's Trident

Percival's Trident Bat.

A Common Sheath-tailed Bat. *JR*

Bat, which is reputed to consume mainly moths. Common Sheath-tailed Bats are much larger (19–41 g) than Percival's Trident Bat (4 g). Like other sheath-tailed bats, Common Sheath-tailed Bats are oral emitters of echolocation calls and, like other trident bats, Percival's Trident Bats are nasal emitters of echolocation calls. Percival's Trident Bats are renowned for the very high frequencies of their echolocation calls, >200 kHz.

Common Sheath-tailed Bats roost in caves and rock shelters. They occur widely across the north of Australia. Percival's Trident Bats roost in caves. This species occurs from coastal Kenya south into South Africa, including Mozambique, Zimbabwe, Zambia and Botswana.

But we have no idea of the population sizes of either species, what threats they may face, or what we could do to conserve them.

So, we know just enough about them to pique our curiosity. Bats' faces epitomize the situation: there is so much more to discover about even this one aspect of their lives and evolution.

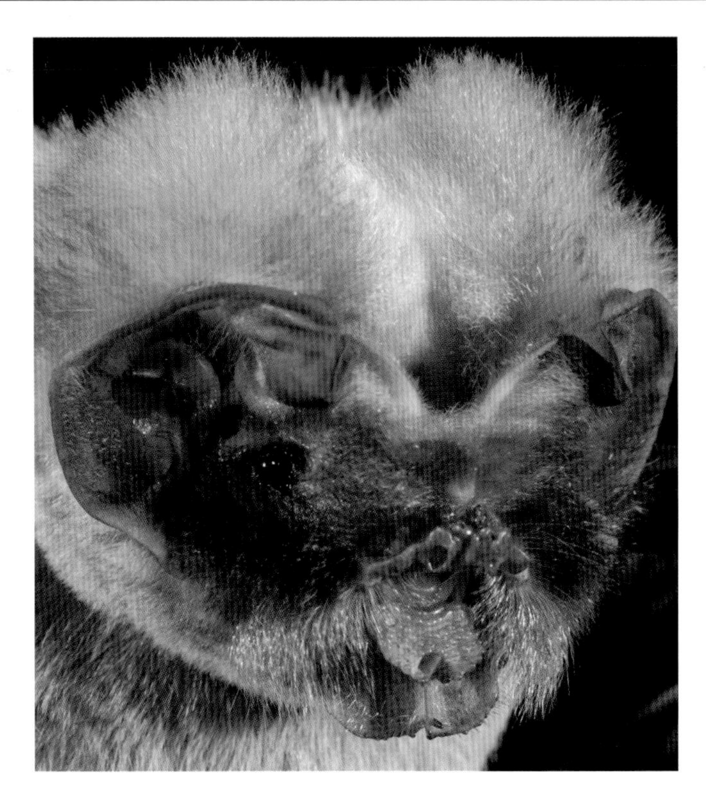

A Wrinkle-faced Bat (above) and a Ghost-faced Bat (below) illustrate the sometimes bizarre faces of these animals. Unravelling the underlying form and function stories behind the faces is a challenge to those who study these alluring animals.

A last word to the bats

On an April evening in Belize in 2022, Brock learned yet another lesson from these endlessly fascinating creatures. We were set up to photograph Black Mastiff Bats emerging from a roost in a tree hollow, where the animals were entering through a long vertical crack. We got ready an hour beforehand, expecting to catch sight of emerging bats as they flew out and away. But they knew we were there. They waited, then walked out of the roost and climbed higher and out of range before flying off. At least one climbed down the tree and walked away. The old adage about never counting chickens can clearly apply to bats too.

Now, Jens could have told us about all this. He would have known and warned us about not disturbing bats you wanted to photograph.

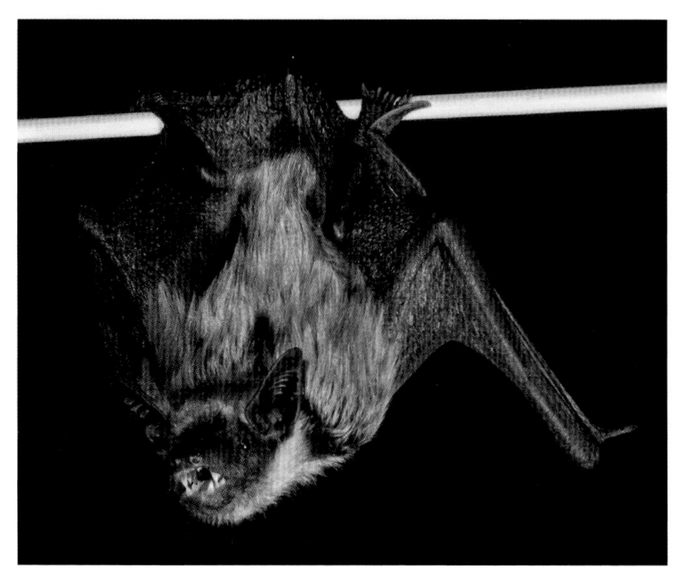

Here are two Big Brown Bat pups. The one on top is six days old, while the pup below is 20 days old. Note the differences. At day 6 its eyes are open, as is the mouth showing the tongue and the tiny milk teeth, but the animal does not yet have fur. By 20 days it is fully furred with erupted permanent teeth and is almost ready to fly. These shots epitomise some of the many things we do not know about bats. How do they learn to fly? To echolocate? Does their mother bring them insect food and supplement it with milk? How long do mother bats stay in contact with their young?

Cast of bats

Common name	Scientific name
Fossil bats	
Small Messel Bat	*Palaeochiropteryx tupaiodon*
Finney's Fossil Bat	*Onychonycteris finneyi*
Pteropodidae: *Old World fruit bats*	
Dyak Fruit Bat	*Dyacopterus spadiceus*
Egyptian Fruit Bat	*Rousettus aegyptiacus*
Geoffroy's Rousette	*Rousettus amplexicaudatus*
Giant Golden-crowned Flying Fox	*Acerodon jubatus*
Hammer-headed Bat	*Hypsignathus monstrosus*
Large Flying Fox	*Pteropus vampyrus*
Leschenault's Rousette	*Rousettus leschenaultii*
Lesser Short-nosed Fruit Bat	*Cynopterus brachyotis*
Pacific Flying Fox (Peka)	*Pteropus tonganus*
Samoan Fruit Bat	*Pteropus samoensis*
Straw-coloured Fruit Bat	*Eidolon helvum*
Wahlberg's Epauletted Fruit Bat	*Epomophorus wahlbergi*
Rhinopomatidae: *Mouse-tailed bats*	
Greater Mouse-tailed Bat	*Rhinopoma microphyllum*
Craseonycteridae: *Hog-nosed bats*	
Kitti's Hog-nosed Bat/Bumblebee Bat	*Craseonycteris thonglongyai*
Rhinolophidae: *Horseshoe bats*	
Bushveld Horseshoe Bat	*Rhinolophus simulator*
Decken's Horseshoe Bat	*Rhinolophus deckenii*
Greater Horseshoe Bat	*Rhinolophus ferrumequinum*
Lesser Horseshoe Bat	*Rhinolophus hipposideros*
Meheyli's Horseshoe Bat	*Rhinolophus meheyli*
Rufous Horseshoe Bat	*Rhinolophus rouxii*
Swinny's Horseshoe Bat	*Rhinolophus swinnyi*
Hipposideridae: *Old World leaf-nosed bats*	
Diadem Leaf-nosed Bat	*Hipposideros diadema*
Great Roundleaf Bat	*Hipposideros armiger*

Common name	Scientific name
Indian Roundleaf Bat	*Hipposideros lankadiva*
Schneider's Leaf-nosed Bat	*Hipposideros speoris*
Sundevall's Roundleaf Bat	*Hipposideros caffer*
Rhinonycteridae: *Trident bats*	
Trident Leaf-nosed Bat	*Asellia tridens*
Percival's Trident Bat	*Cloeotis percivali*
Megadermatidae: *False vampire bats*	
Australian Ghost Bat	*Macroderma gigas*
Greater False Vampire Bat	*Megaderma lyra*
Heart-nosed Bat	*Cardioderma cor*
Lesser False Vampire Bat	*Megaderma spasma*
Yellow-winged Bat	*Lavia frons*
Nycteridae: *Slit-faced bats*	
Egyptian Slit-faced Bat	*Nycteris thebaica*
Large Slit-faced Bat	*Nycteris grandis*
Emballonuridae: *Sheath-tailed bats*	
Common Sheath-tailed Bat	*Taphozous georgianus*
Greater Sac-winged Bat	*Saccopteryx bilineata*
Hildegard's Tomb Bat	*Taphozous hildegardeae*
Lesser Dog-like Bat	*Peropteryx kappleri*
Pale-winged Dog-like Bat	*Peropteryx pallidoptera*
Proboscis Bat	*Rhynchonycteris naso*
Phyllostomidae: *New World leaf-nosed bats*	
Antillean Fruit-eating Bat	*Brachyphylla cavernarum*
Big-eared Woolly Bat	*Chrotopterus auritus*
Buffy Flower Bat	*Erophylla sezekorni*
California Leaf-nosed Bat	*Macrotus californicus*
Common Tent-making Bat	*Uroderma bilobatum*
Common Vampire Bat	*Desmodus rotundus*
Davis's Round-eared Bat	*Lophostoma evotis*
Fringe-lipped Bat	*Trachops cirrhosus*

Common name	Scientific name
Golden Bat	*Mimon cozumelae*
Great Fruit-eating Bat	*Artibeus lituratus*
Greater Spear-nosed Bat	*Phyllostomus hastatus*
Heller's Broad-nosed Bat	*Platyrrhinus helleri*
Hairy-legged Vampire Bat	*Diphylla ecaudata*
Honduran White Bat	*Ectophylla alba*
Jamaican Fruit Bat	*Artibeus jamaicensis*
Lesser Long-nosed Bat	*Leptonycteris yerbabuenae*
Mexican Long-tongued Bat	*Choeronycteris mexicana*
Northern Yellow-shouldered Bat	*Sturnira parvidens*
Pale Spear-nosed Bat	*Phyllostomus discolor*
Pallas's Long-tongued Bat	*Glossophaga soricina*
Pygmy Fruit-eating Bat	*Artibeus phaeotis*
Seba's Short-tailed Fruit Bat	*Carollia perspicillata*
Spectral Bat	*Vampyrum spectrum*
Striped Hairy-nosed Bat	*Gardnerycteris keenani*
Thomas's Fruit-eating Bat	*Dermanura watsoni*
White-winged Vampire Bat	*Diaemus youngi*
Wrinkle-faced Bat	*Centurio senex*
Yellow-eared bats	*Vampyressa* spp.
Yellow-throated Big-eared Bat	*Lampronycteris brachyotis*
Mormoopidae: *Moustached bats*	
Antillean Ghost-faced Bat	*Mormoops blainvillei*
Big Naked-backed Bat	*Pteronotus gymnonotus*
Davy's Naked-backed Bat	*Pteronotus fulvus*
Ghost-faced Bat	*Mormoops megalophylla*
Lesser Moustached Bat	*Pteronotus psilotis*
Mesoamerican Moustached Bat	*Pteronotus mesoamericanus*
Parnell's Moustached Bat	*Pteronotus parnellii*
Sooty Moustached Bat	*Pteronotus quadridens*

Common name	Scientific name
Noctilionidae: *Bulldog bats*	
Greater Bulldog Bat	*Noctilio leporinus*
Mystacinidae: *Short-tailed bats*	
Lesser Short-tailed Bat	*Mystacina tuberculata*
Thyropteridae: *New World disk-winged bats*	
Spix's Disk-winged Bat	*Thyroptera tricolor*
Natalidae: *Funnel-eared bats*	
Gervais's Funnel-eared Bat	*Nyctiellus lepidus*
Mexican Funnel-eared Bat	*Natalus mexicanus*
Molossidae: *Free-tailed bats*	
Angolan Free-tailed Bat	*Mops condylurus*
Black Mastiff Bat	*Molossus nigricans*
Brazilian Free-tailed Bat	*Tadarida brasiliensis*
Dwarf Bonneted Bat	*Eumops nanus*
European Free-tailed Bat	*Tadarida teniotis*
Florida Bonneted Bat	*Eumops floridanus*
Large-eared Free-tailed Bat	*Otomops martiensseni*
Naked Bat	*Cheiromeles torquatus*
Roberts's Flat-headed Bat	*Sauromys petrophilus*
Velvety Free-tailed Bat	*Molossus molossus*
Miniopteridae: *Bent-winged bats*	
Greater Long-fingered Bat	*Miniopterus inflatus*
Lesser Long-fingered Bat	*Miniopterus fraterculus*
Natal Long-fingered Bat	*Miniopterus natalensis*
Schreiber's Long-fingered Bat	*Miniopterus schreibersii*
Vespertilionidae: *Vesper bats*	
Argentinian Brown Bat	*Eptesicus furinalis*
Banana Pipistrelle	*Neoromicia nanus*
Bechstein's Bat	*Myotis bechsteinii*
Big Brown Bat	*Eptesicus fuscus*
Bird-like Noctule	*Nyctalus aviator*

Common name	Scientific name
Bodenheimer's Pipistrelle	*Hypsugo bodenheimeri*
Botswana Long-eared Bat	*Laephotis botswanae*
Brandt's Bat	*Myotis brandtii*
Brown Long-eared Bat	*Plecotus auritus*
California Myotis	*Myotis californicus*
Cape Hairy Bat	*Myotis tricolor*
Common Noctule	*Nyctalus noctula*
Common Pipistrelle	*Pipistrellus pipistrellus*
Daubenton's Bat	*Myotis daubentonii*
Desert Pipistrelle	*Hypsugo ariel*
Dusky Pipistrelle	*Pipistrellus hesperidus*
Eastern Small-footed Bat	*Myotis leibii*
Eastern Red Bat	*Lasiurus borealis*
Fishing Myotis	*Myotis vivesi*
Fringed Myotis	*Myotis thysanodes*
Greater Evening Bat	*Ia io*
Greater Mouse-eared Bat	*Myotis myotis*
Greater Noctule	*Nyctalus lasiopterus*
Harlequin Bat	*Scotomanes ornatus*
Hoary Bat	*Lasiurus cinereus*
Japanese House Bat	*Pipistrellus abramus*
Kuhl's Pipistrelle	*Pipistrellus kuhlii*
Lesser Noctule/Leisler's Bat	*Nyctalus leisleri*
Little Brown Myotis	*Myotis lucifugus*
Nathusius' Pipistrelle	*Pipistrellus nathusii*
Natterer's Bat	*Myotis nattereri*
Noctules	*Nyctalus* spp.
Northern Bat	*Eptesicus nilssonii*
Northern Long-eared Bat	*Myotis septentrionalis*
Painted Bat	*Kerivoula picta*
Pallid Bat	*Antrozous pallidus*

Common name	Scientific name
Particoloured Bat	*Vespertilio murinus*
Rickett's Big-Footed Bat	*Myotis pilosus*
Round-eared Tube-nosed Bat	*Murina cyclotis*
Savi's Pipistrelle	*Pipistrellus savii*
Serotine	*Eptesicus serotinus*
Silver-haired Bat	*Lasionycteris noctivagans*
Silver-tipped Myotis	*Myotis albescens*
Soprano Pipistrelle	*Pipistrellus pygmaeus*
Spotted Bat	*Euderma maculatum*
Tricoloured Bat	*Perimyotis subflavus*
Ussuri Tube-nosed Bat	*Murina ussuriensis*
Van Gelder's Bat	*Bauerus dubiaquercus*
Welwitsch's Bat	*Myotis welwitschii*
Western Barbastelle	*Barbastella barbastellus*
Yuma Myotis	*Myotis yumanensis*
Zulu Serotine	*Neoromicia zuluensis*

Notes

Chapter 1

[1] Fenton, M.B. 1975. Observations on the biology of some Rhodesian bats including a key to the Chiroptera of Rhodesia. *Life Science Contributions, Royal Ontario Museum* 104: 1–27.

[2] Jones, T.K. and C.F. Moss. 2021. Visual cues enhance obstacle avoidance in echolocating bats. *Journal of Experimental Biology* 224. DOI: 10.1242/jeb.242968

[3] Salles, A. 2022. Bats: vision or echolocation, why not both? *Current Biology* 32: R311–R333.

[4] Podlutsky, A., N.D. Ovodov and S.V. Austad. 2005. A new field record for bat longevity. *Journal of Gerontology* 60A: 1366–8.

[5] Rodriguez, H.V. and C.A. Mancina. 2020. Alta longevidad en el murciélago mariposa, *Nyctiellus Lepidus* (Gervais, 1837) (Natalidae), uno de los murciélagos más pegueños de mundo. *Journal of Bat Research and Conservation*. DOI: 10.14709/Barbj.13.1.2020.16

[6] Wilkinson, G.S. and D.M. Adams. 2018. Recurrent evolution of extreme longevity in bats. *Biology Letters*. http//dx.doi.org/10.1098/rsbl.2018.0860

[7] Alcade, J.T., M. Jiminez, I. Brilla, V. Vintullis, C.C. Voigt and G. Petersons. 2020. Transcontinental 2200 km migration of a Nathusius' pipistrelle (*Pipistrellus nathusii*) across Europe. *Mammalia*. https://doi.org/10.1515/mammalia-2020-0069

[8] Fenton, M.B. 1969. Summer activity of *Myotis lucifugus* (Chiroptera: Vespertilionidae) at hibernacula in Ontario and Quebec. *Canadian Journal of Zoology* 47: 597–602.

Chapter 2

[9] Simmons, N.B., K.L. Seymour, J. Habersetzer and G.F. Gunnell. 2008. Primitive Early Eocene bat from Wyoming and the evolution of flight and echolocation. *Nature*. DOI: 10.1038/nature06549

[10] Sterbing-D'Angelo, M. Chadha, C. Chiu, B. Falk, W. Xian, J. Barcelo, J.M. Zook, and C.F. Moss. 2011. Bat wing sensors support flight control. *Proceedings of the National Academy of Sciences of the USA*. www.pnas.org/cgi/doi/10.1073/pnas.1018740108

[11] Swartz, S.M. and N. Konow. 2015. Advances in the study of bat flight: the wing and the wind. *Canadian Journal of Zoology*. DOI: 10.1139/cjz-2015-0117

[12] Meierhofer, M.B. , J.S. Johnson, K.A. Field, S.S. Lumadue, A. Kurta, J.A. Kath and D.M. Reeder. 2018. Bats recovering from white-nose syndrome elevate metabolic rate during wing healing in spring. *Journal of Wildlife Diseases* 54: 480–90.

[13] Rummel, A.D., S.M. Swartz and R.L. Marsh. 2019. Warm bodies, cool wings: reginal heterothermy in flying bats. *Journal of Experimental Biology*. http://dx.doi.org/10.1098.rsbl.2019.0530

[14] Rydell, J., M.B. Fenton, E. Seamark, P.W. Webala and T.C. Michaelsen. 2020. White and clear wings in bats (Chiroptera). *Canadian Journal of Zoology*. DOI: 10.1139/cjz-2019-0182

[15] Aldridge, H.D.J.N. 1988. Flight kinematics and energetics in the little brown bat, *Myotis lucifugus* (Chiroptera: Vespertilionidae), with reference to the influence of ground effect. *Journal of Zoology London* 216: 5–7.

[16] McCracken, G.F., K. Safi, T.H. Kunz, D.K.N. Dechmann, S.M. Swartz and M. Wikelski. 2016. Airplane tracking documents the fastest flight speeds recorded for bats. *Royal Society Open Science*. https://doi.org/10.1098/rsos.160393

[17] O'Mara, M.T., F. Amorim, M. Scacco and G.F. McCracken. 2021. Bats use topography and nocturnal updrafts to fly high and fast. *Current Biology* 31: 1311–16. https://doi.org/10.1016/jcub.2020.12.024

Chapter 3

[18] He, K., Q. Liu, D-M. Xu, F-Y. Qi, J. Bai, S-W. He, P. Chen, X. Zhou, W-Z. Cai, Z-Z. Chen. Et al. 2021. Echolocation in soft-furred tree mice. *Science* 372: e513.

[19] Thaler, L., G.M. Reich, X. Zhang, D. Wang, G.E. Smith, Z. Tao, R., R.S.M.A. Bin, R. Abdullah, M. Cherniakov, C.J. Baker, D. Kish and M. Antoniou. 2017. Mouth-clicks used by blind expert human echolocators – signal description and model based signal synthesis. *PLoS Computational Biology* 3(8):e1005670. https://doi.org/10.1371/journal.pcbi.1005670

[20] Barclay, R.M.R., J.H. Fullard and D.S. Jacobs. 1999. Variation in the echolocation calls of the hoary bat (*Lasiurus cinereus*): influence of body size, habitat structure, and geographic location. *Canadian Journal of Zoology* 77: 530–34.

[21] Fenton, M.B. 2013. Questions, ideas, and tools: lessons from bat echolocation. Anniversary Essay. *Animal Behaviour*. http://dx.doi.org/10.1016/j.anbehav.2013.02.024

[22] Veselka, N., D.D. McErlain, D.W. Holdsworth, J.L. Eger, R.K Chhem, M.J. Mason, K.LO Brain, P.A. Faure and M.B. Fenton. 2010. A bony connection signals laryngeal echolocation in bats. *Nature*. DOI: 10.1038/nature08737

[23] Sulser, R.B., B.D. Patterson, D.J. Urban, A.I. Neander and Z-X Luo. 2022. Evolution of inner ear neuroanatomy of bats and implications for echolocation. *Nature*. https://doi.org/101038/s41586-021-04335-z

[24] Curtis, A.A., J.H. Arbour and S.E. Santana. 2020. Mind the gap: natural cleft palates reduce biting performance in bats. *Journal of Experimental Biology* jeb196535. DOI: 10.1242/jeb.196535

[25] Obrist, M.K., M.B. Fenton, J.L. Eger and P. Schlegel. 1993. What ears do for bats: a comparative study of pinna sound pressure transformation in Chiroptera. *Journal of Experimental Biology* 80: 119–52.

[26] Håkansson, L. Jakobsen, A. Hedenström, and L.C. Johansson. 2017. Body lift, drag and power are relatively higher in large-eared than in small-eared species. *Journal of the Royal Society Interface*. https://doi.org/10.1098/rsif.2017.0455

[27] Keeley, B.W., A.T.H. Kelley and P. Houlahan. 2018. Ridge number in bat ears is related to both guild membership and ear length. *PLoS ONE* https://doi.org/10.1371/journal.pone.0200255

Chapter 4

[28] Corcoran, A.J. and T.J. Weller. 2018. Inconspicuous echolocation in hoary bats (*Lasiurus cinereus*). *Proceedings B Royal Society*. http://dx/doi.org/10.1098/rspb.2018.0441

[29] Bohn, K.M. and E.H. Gillam. 2018. In-flight social calls: a primer for biologists and managers studying echolocation. *Canadian Journal of Zoology* 96: 787–800.

[30] Beetz, M.J., M. Kössl, and J.C. Hechavarria. 2021. The frugivorous bat *Carollia perspicillata* dynamically changes echolocation parameters in response to acoustic playback. *Journal of Experimental Biology* 224: jeb234245. DOI: 10.1242/jeb.234245

[31] Acharya, L. and M.B. Fenton. 1999. Bat attacks and moth defensive behaviour around street lights. *Canadian Journal of Zoology* 77: 27–33.

[32] Goerlitz, H.R., H.M. ter Hofstede, M.A.R.K. Zeale and G. Jones. 2010. An aerial-hawking bat uses stealth echolocation to counter moth hearing. *Current Biology*. DOI: 10.1016/j.cub.2010.07.046

[33] Neil, T.R., Z. Shen, D. Robert, B.W. Drinkwater and M.W. Holdereid. 2020. Thoracic scales of moths as a stealth coating against bat biosonar. *Journal of the Royal Society Interface* 17: 20190692.

[34] Corcoran, J.J. Barber, N.I. Hristov, and W.E. Connor. 2011. How do tiger moths jam bat sonar? *Journal of Experimental Biology* 214: 2416–25.

[35] Chaverri, G., N.I. Sandoval-Herrera, P. Iturralde-Polit, A. Romero-Vasquez, S.Chaves-Ramirez and M. Sagot. 2021. The energetics of social signaling during roost location in Spix's disk-winged bats. *Journal of Experimental Biology*. DOI: 10.1242/jeb.238279

[36] Barclay, R.M.R., M.B. Fenton, M.D. Tuttle, and M.J. Ryan. 1981. Echolocation calls produced by *Trachops cirrhosus* (Chiroptera: Phyllostomatidae) while hunting for frogs. *Canadian Journal of Zoology* 59: 750–3.

[37] Ratcliffe, J.M., H. Ragnhuram, G. Marimuthu, J.H. Fullard and M.B. Fenton. 2005. Hunting in unfamiliar space: echolocation in the Indian false vampire bat, *Megaderma lyra. Behavioural Ecology and Sociobiology* 58: 157–64.

[38] Prakash, H., S. Greif, Y. Yovel and R. Balakrishnan. 2021. Acoustically eavesdropping bat predators take longer to capture katydid prey signaling in aggregation. *Journal of Experimental Biology*. DOI: 10.1242/jeb.233262

Chapter 5

[39] Ingala, M.R., N.B. Simmons and S.L. Perkins. 2018. Bats are an untapped system for understanding microbiome evolution in mammals. *Msphere*. https//doi.org/10.1128/msphere.00397-18; Ingala, M.R., N.B. Simmons, C.Wultsch, K. Krampis, K.L. Provost and S.L. Perkins. Molecular diet analysis of neotropical bats based on fecal DNA metabarcoding. *Ecology and Evolution*. DOI: 10.1002/ece.7579

[40] De la Cueva, H., M.B. Fenton, M.B.C. Hickey, and R.W. Baker. 1995. Energetic consequences of flight speeds of foraging red and hoary bats (*Lasiurus borealis* and *Lasiurus cinereus*; Chiroptera: Vespertilionidae). *Journal of Experimental Biology* 198: 2245–51.

[41] Fenton, M.B., I.L. Rautenbach, J. Rydell, H.W. Arita, J. Orgega, S. Bouchard, M.D. Hovorka, B.K. Lim, E. Odgren, C.V. Portfors-Yeomans, W. Scully, D.M. Syme and M.J. Vonhof. 1998. Emergence, echolocation, diet and foraging of *Molossus ater. Biotropica* 30: 314–20.

[42] Barclay, R.M.R. and R.M. Brigham. 1994. Constraints on optimal foraging: a field test of prey discrimination by echolocating insectivorous bats. *Animal Behaviour* 48: 1013–21.

[43] Ibáñez, C., J. Juste, J.L. García-Mudarra, and P.T. Agirre-Mendi. 2001. Bat predation on nocturnally migrating birds. *Proceedings of the National Academy of Sciences of the USA*. https://pnas.org/cgi/doi/10.1073/pnas.171140598

[44] Norberg, U.M. & M.B. Fenton.1988. Carnivorous bats? *Biological Journal of the Linnean Society* 33: 383–94.

[45] Clare, E.L., W.O.C. Symondson, H. Broders, F. Fabianek, E.E. Fraser, A. MacKenzie, A. Boughen, R. Hamilton, C.K.R. Willis, F. Martinez-Nunez, A.K. Menzies, K.J.O. Norquay, M. Brigham, J. Poissant, J. Rintoul, R.M.R. Barclay and J.P. Reimer. 2113. The diet of *Myotis lucifugus* across Canada: assessing foraging quality and diet variability. *Molecular Ecology*. DOI: 10.111/mec.12542

[46] Fujioka, E., I. Aihara, M. Sumiya, K. Aihara and S. Hiryu. 2016. Echolocating bats use future target information for optimal foraging. *Proceedings of the National Academy of Sciences of the USA*. www.pnas.org//cgi/doi/10.1073/pnas.151509113

[47] Salles, A., C.A. Diebold and C.F. Moss. 2020. Echolocating bats accumulate information from acoustic snapshots to predict auditory object motion. *Proceedings of the National Academy of Sciences of the USA*. www.pnas.org/cgi/doi/101073/pnas.2011719117

[48] Boonman, A., M.B. Fenton and Y. Yovel. 2019. The benefits of insect-swarm hunting to echolocating bats, and its influence on the evolution of bat echolocation signals. *PLoS Computational Biology* 15(12): e1006873. https://doi.org/10.1371/journal.pcbi.1006873

[49] Johansson, L.C., L. Jakobsen and A. Hedenström. 2018. Flight in ground effect dramatically reduces aerodynamic costs in bats. *Current Biology*. https://doi.org/10.1016/j.cub.2018.09.011

[50] Rodriguez-Duran, A, and J. Rosa. 2020. Remarkable variation in the diet of Noctilio leporinus in Puerto Rico: the fishing bat turns carnivorous. *Acta Chiropterologica*. https://doi.org/10.3161/15081109AACC2020.22.1.016

[51] Ghose, R.K. and D.K. Ghosal. 1984. Record of the fulvous fruit bat, *Rousettus leschenaultia* (Desmarest, 1820) from Sikkim, with notes on its interesting feeding habit and status. *Journal of the Bombay Natural History Society*. https://www.biodiversitylibrary.org/bibliography/7414

Chapter 6

[52] Clare, E.L., H.R. Goerlitz, V.A. Drapeau, M.W. Holderied, A.M. Adams, J. Nagel, E.R. Dumont, P.D.N. Hebert and M.B. Fenton. 2013. Trophic niche flexibility in *Glossophaga*

soricina: how a nectar seeker sneaks an insect snack. *Functional Ecology* 28: 632–41. DOI: 10.1111/1365-2435.12192

[53] Brokaw, A.F. and M. Smotherman. 2021. Olfactory tracking strategies in a neotropical fruit bat. *Journal of Experimental Biology*. DOI: 10.1242/jeb231829

[54] Fleming, T.H., C. Geiselman and W.J. Kress. 2009. The evolution of bat pollination: a phylogenetic perspective. *Annals of Botany* 104: 1017–43.

[55] Datzmann, T., O. von Helversen and F. Mayer. 2010. Evolution of nectarivory in phyllostomid bats (Phyllostomidae Gray, 1825, Chiroptera: Mammalia). BMC *Evolutionary Biology*. http://www.biomedcentral.com/`471-2148/10/165

[56] Ingersoll, R., L. Haizmann and D. Lentink. 2018. Biomechanics of hover performance in Neotropical hummingbirds versus bats. *Science Advances* 4: 2980.

[57] Vejdani, H.R., D.B. Boerma, S.M. Swartz and K.S. Breuer. 2019. The dynamics of hovering flight in hummingbirds, insects and bats with implications for aerial robotics. *Bioinspiration and Biomimetics* 14: 0166003.

[58] Harper, C.J., S.M. Swartz and E.L. Brainerd. 2013. Specialized bat tongue is a hemodynamic nectar mop. *Proceedings of the National Academy of Sciences of the USA* 110: 8852–7.

[59] Fleming, T.H., S.L. Fenton and M.B. Fenton. 2020. Hummingbird and bat pollinators of the Chiricahuas. *American Scientist* 108(6): 362.

[60] Murphy, M., E.L. Clare, J. Rydell, Y. Yovel, Y. Bar-On, P. Olebaum and M.B. Fenton. 2015. The opportunistic use of banana bracts by *Glossophaga soricina* in Belize. *Acta Chiropterologica* 18: 209–13.

[61] Alpizar, P., A. Risely, M. Tschapka and S. Sommer. 2021. Agricultural fast food: bats feeding in banana monocultures are heavier but have less diverse gut microbiota. *Frontiers in Ecology and Evolution* 9: 746783.

Chapter 7

[62] Carter, G.G., D.R. Farine, R.J. Crisp, J.K. Vrtilek, S.P. Ripperger and R.A. Page. 2020. Development of food-sharing relationships in vampire bats. *Current Biology*. https://doi.org/10.1016/j.cub.2020.01.055

[63] Evans, D.L., B.N. Vis, N.P. Dunning, E. Graham and C. Isendahl. 2021. Buried solutions: how Maya urban life substantiates soil connectivity. *Geoderma* 387: 114925. https://doi.org/10.1016/j.geoderma.2020.114925

[64] Grotta-Neto, F., P.H. de Faria Peres, U. Piovezan and F. Passos. 2021. Hunting practices of feral pigs (*Sus scrofa*) and predation by vampire bats (*Desmodus rotundus*) as a potential route of rabies in the Brazilian Pantanal. *Animal Ecology*. DOI: 10.1111/aec.12971

[65] Blumer, M., T. Brown, M. B. Freitas, A.L. Destro, J.A. Oliveira, A.E. Morales, T. Schell, C. Greve, P. Pippel, D. Jebb, N. Hecker, A-W. Ahmed, B.M. Kirilenko, M. Foote, A. Jenka, B.K. Lim and M. Hiller. 2022. Gene loses in the common vampire bat illuminate molecular adaptations to blood-feeding. *Science Advances* 8: eabm6494.

Chapter 8

[66] Rydell, J., D. Russo, P. Sewell, E.C.J. Seamark, C.M. Frances, S.L. Fenton and M.B. Fenton. 2022. Bat selfies – photographic surveys of flying bats. *Mammalian Biology*. https://doi.org/10.1007/s42991-022-00233-7

[67] Audet, D. and M.B. Fenton. 1988. Heterothermy and the use of torpor by the bat *Eptesicus fuscus* (Chiroptera: Vespertilionidae): a field study. *Physiological Zoology* 61: 197–204.

[68] Barclay, R.M.R., D.S. Jacobs, C.T. Harding, A.E. McKechnie, S.D. McCulloch, W. Markotter, J. Pawseka, and R.M. Brigham. 2017. Thermoregulation by captive and free-ranging Egyptian rousette bats (*Rousettus aegyptiacus*) in South Africa. *Journal of Mammalogy*. https://doi.org/101093/jmammal/gyw234

[69] Willis, C.K.R. and R.M. Brigham. 2004. Roost switching, roost sharing and social cohesion: forest-dwelling big brown bats, *Eptesicus fuscus*, conform to the fission–fusion model. *Animal Behaviour*. DOI: 10.1016/j.anbehav.2003.08.028
[70] Willis, C.K.R. and R.M. Brigham. 2005. Physiological and ecological aspects of roost selection by reproductive female hoary bats (*Lasiurus cinereus*). *Journal of Mammalogy* 86: 85–94.
[71] Czenze, Z.J. and M.B. Dunbar. 2017. Hot bats go cold: heterothermy in neotropical bats. *Canadian Journal of Zoology*. https://doi.org/10.1139/cjz-2016-0318
[72] Herrera M, G.L., V.B. Salinas-Ramos, J.J. Florese Martinez and D. Johnston. 2014. Winter and summer torpor in a free-ranging subtropical desert bat: the fishing myotis (*Myotis vivesi*). *Acta Chiropterologica*. DOI: 10.3161/150811014X687288
[73] Hirakawa, H. and Y. Nagasaka. 2018. Evidence for Ussurian tube-nosed bats (*Murina ussuriensis*) hibernating in snow. *Scientific Reports*: https://doi.org/10.1038/s41598-018-30357-1
[74] Cholewa, E., M.J. Vonhof, S. Bouchard, C.A. Peterson and M.B. Fenton. 2001. The pathway of water movement in leaves modified into tents by bats. *Biological Journal of the Linnean Society* 72: 179–91.
[75] Bondo, K.J., C.K.R. Willis, J.D. Metheny, R.J. Kilgour, E.H. Gillam, M.C. Kalcounis-Rueppell and R.M. Brigham. 2019. Bats relocate maternity colony after the natural loss of roost trees. *Journal of Wildlife Management*. DOI: 10.1002/jwmg.21751
[76] Rydell, J. 1992. Occurrence of bats in northernmost Sweden (65oN) and their feeding ecology in summer. *Journal of Zoology* 227: 517–29.
[77] Rydell, J., K-B. Strann and J.R. Speakman. 1994. First record of breeding bats above the Arctic Circle: Northern bats at 68–70oN in Norway. *Journal of Zoology* 233: 335–9.
[78] Hernández-Montero, J., C. Reusch R. Simon, C.R. Schöner and G. Kerth. 2020. Free-ranging bats combine three different cognitive processes for roost localization. *Oecologia* 292: 979–88.

Chapter 9

[79] Bradbury, J.W. 1977. Lek mating behaviour in the hammer-headed bat. *Z. Tierpsychologie* 45: 225–55.
[80] Urban, D.J., D.W. Sorensen, J.A. Maier, M.B. Fenton, N.B. Simmons, L.N. Cooper and K.E. Sears. 2015. Conjoined twins in a wild bat: a case report. *Acta Chiropterologica*. DOI: 10.3161/150811o9AC2015.7.1.016
[81] Vaughan, T.A. and R.P. Vaughan. 1987. Parental behaviour in the African yellow-winged bat (*Lavia frons*). *Journal of Mammalogy* 68: 217–23
[82] Goldshtein, A., L. Harten and Y. Yovel. 2022. Mother bats facilitate pup navigation learning. *Current Biology* 32: 1–11. https://doi.org/10.1016/j.cub.2021.11.010
[83] Faulkes, C.G., J.S. Elmore, D.A. Baines, M.B. Fenton, N.B. Simmons and E.L. Clare. 2019. Chemical characterisation of potential pheromones from the shoulder gland of the Northern yellow-shouldered-bat, *Sturnira parvidens* (Phyllostomidae: Stenodermatinae). *PeerJ* 7: e7734. DOI: 10.7717/peerj.7734
[84] Bell, G.P. 1980. Habitat use and response to patches of prey by desert insectivorous bats. *Canadian Journal of Zoology* 58: 1876–83.
[85] Egert-Berg, K., M. Handel, A. Goldstein, O. Eitan, I. Bokrissov and Y. Yovel. 2018. Resource ephemerality drives social foraging in bats. *Current Biology*. https://doi.org/10.1016/j.cub.2018.09.064
[86] Walter, M.H., A. Verdon, V. Olmos, C.C. Weiss, L-R. Vial, A. Putra, J. Muller, M. Tschapka and H-U. Schnitzler. 2020. Discrimination of small sugar concentration differences helps the nectar-feeding bat *Leptonycteris yerabuenae* cover energetic demands. *Journal of Experimental Biology* 223: jeb215053. DOI:10.1242/jeb.215053
[87] Gaudet, C.L. and M.B. Fenton. 1984. Observational learning in three species of insectivorous bats (Chiroptera). *Animal Behaviour* 32: 385–8.

Chapter 10

[88] Davis, W.H. and H.B. Hitchcock 1965. Biology and migration of the bat, *Myotis lucifugus*, in New England. *Journal of Mammalogy* 46: 296–313.

[89] Williams, T.C., J.M. Williams and D.R. Griffin. 1966. The homing ability of the neotropical bat *Phyllostomus hastatus*, with evidence for visual orientation. *Animal Behaviour* 14: 468–73.

[90] Roberts, B.J., C.P. Catterall, P. Eby and J. Kanowski. 2012. Long-distance and frequent movements of the flying fox *Pteropus poliocephalus*: implications for management. *PLoS ONE*. DOI: 10.1371/journal.pone.0042532

[91] McGuire, L.P., K. Jonasson and C.G. Gugliemo. 2011. Migratory stopover in the long distance migrant silver-haired bat, *Lasionycteris noctivagans*. *Journal of Animal Ecology* https://doi.org/10.1111/j.1365-2656.2011.01912.x

[92] Thorne, T.J., E. Matczak, M. Donnelly, M.C. Franke and K.C.R. Kerr. 2021. Occurrence of a forest-dwelling bat, northern myotis (*Myotis septentrionalis*), within Canada's largest conurbation. *Journal of Urban Ecology*. DOI: 10.1093/jue/juab029

[93] Toldeo, S., D. Schohaml, I. Schiffner, E. Lourie, Y. Orchan, Y. Bartan and R. Nathan. 2020. Cognitive map-based navigation in wild bats revealed by a new high-throughput tracking system. *Science*. DOI: 10:1126/science.aax6904

[94] Harten, L., A. Katz, A. Goldshtein, M. Handel and Y. Yovel. 2020. The ontogeny of a mammalian cognitive map in the real world. *Science* 369: 194–7. DOI: 10.1126/science.aay3354

[95] Dotson, N.M. and M.M. Yartsev. 2021. Nonlocal spatiotemporal representation in the hippocampus of freely flying bats. *Science*: http://science.sciencemag.org/101126/science.abg1278

[96] Burns, L.E. and H.G. Broders. 2015. Who swarms with whom? Group dynamics of Myotis bats during autumn swarming. *Behavioural Ecology*. DOI: 10.1093/beheco/arv017; Burns, L.E. and H.G. Broders. 2015. Maximizing mating opportunities: higher autumn swarming activity in male versus female *Myotis bats*. *Journal of Mammalogy*. DOI: 10.1093/jmammal/gyv141

Chapter 11

[97] Fenton, M.B., I.L. Rautenbach, S.E. Smith, C.M. Swanepoel, J. Grosell and J. van Jaarsveld. 1994. Raptors and bats: threats and opportunities. *Animal Behaviour* 48: 9–18.

[98] Jones, G. and J. Rydell. 1994. Foraging strategy and predation risk as factors influencing emergence time in echolocating bats. *Philosophical Transactions of the Royal Society* B 346: 445–55.

[99] Boinski, S. and R.M. Timm. 1985. Predation by squirrel monkeys and double-toothed kites on tent-making bats. *American Journal of Primatology* 9: 121–7.

[100] Estók, P., S. Zsebők and B.M. Siemers. 2010. Great tits search for, capture, kill and eat hibernating bats. Biology Letters 6(1): 59–62. https://doi.org/10.1098/rsbl.2009.0611

[101] Marshall, A.G. 1971. The ecology of Basilia hispida (Diptera: Nycteribiidae) in Malaysia. *Journal of Animal Ecology* 40: 141–54.

[102] Talbot, N., N. Keyghobadi and M.B. Fenton. 2018. Bed bugs: the move to humans as hosts. *FACETS* 4: 105–10.

[103] Speer, K.A. 2021. Microbiomes mediate host–parasite interactions. *Molecular Ecology News and Views*. DOI: 10.1111/mec.16381

[104] Cryan, P.M., P.M. Gorresen, C.D. Hein, M.R. Schirmacher, R.H. Diehl, M.M. Huso, D.T.S. Hayman, P.D. Fricker, F.J. Bonaccorso, D.H. Johnson, K. Heist and D.C. Dalton. 2014. Behaviour of bats at wind turbines. *Proceedings of the National Academy of Sciences of the USA* 111: 15126–31. https://doi.org/10.1073/pnas.1406672111

[105] Frick, W. T. O'Mara, M. Wikelski, B. Kranstauber and D.K.N. Dechmann. 2017. Fatalities at wind turbines may threaten population viability of a migratory bat. *Biological Conservation* 209: 172–7. https://doi.org/10.1016/j.biocon.2017.02.023

106 Jansson, S., E. Malmqvist, M. Brydegaard, S. Akesson and J. Rydel. 2020. Insect swarm dynamics at a wind turbine observed with Scheimpflug lidar. *Ecological Indicators* 117: 106578. https://doi.org/10.1016/j.ecolind.2020.106578

107 Cornman. R.S., J.A. Fike, S.J. Oyler-McCance and P.M. Cryan. 2020. Historical effective population size of North American hoary bat (*Lasiurus cinereus*) and challenges to estimating trends in contemporary effective breeding population size from archived samples. *PeerJ*. DOI: 10.7717/peerj.11285

108 Rydell, J., J. Eklöf, and S. Sánchez-Navarro. 2017. Age of enlightenment: long-term effects of outdoor aesthetic lights on bats in churches. *Royal Society Open Science* 4: 161077.

109 Frick, W.F., J.F. Pollock, A.C. Hicks, K.E. Langwig, D.S. Reynolds, G.G. Turner, C.M. Butchkoski and T.H. Kunz. 2010. An emerging disease causes regional population collapse of a common North American bat species. *Science*. DOI: 10.1126/science.1188594

110 Dzal, Y., L.P. McGuire, N. Veslka and M.B. Fenton. 2011. Going, going, gone: the impact of white-nose syndrome on the summer activity of the little brown bat (*Myotis lucifugus*). *Biology Letters* 7: 392–4. DOI: 10.1098/rsbl.2010.0859

111 Rebelo, H., P. Tarroso and G. Jones. 2010. Predicted impact of climate change on European bats in relation to their biogeographic patterns. *Global Change Biology* 16: 561–76.

112 Roche, N., S.D. Langton, T. Aughney and D. Lynn. 2019. Elucidating the consequences of a warming climate for common bat species in north-western Europe. *Acta Chiropterologica* 21(2): 359–73. DOI: 10.3161/15081109ACC2019.21.2.011

113 Rydell, J., M. Elfström, J. Eklöf and S. Sånchez-Navarro. 2020. Dramatic decline of northern bat *Eptesicus nilssonii* in Sweden over 30 years. *Royal Society Open Science* 7: 191754; De Bruyn, L., R. Gyselings, L. Kirkpatrick and A. Rachwald. 2020. Temperature driven hibernation site use in the western barbastelle *Barbastella barbastellus* (Schreber, 1774). *Scientific Reports*. https://doi.org/10.1038/s41598-020-80720-4; Meij, T. Van der, A.J. Van Stsrien, K.A. Haysom, J. Dekker, J. Russ, K. Biala, Z. Bihari, E. Jansen, S. Langton, A. Kurali, H. Limpens, A. Meschede, G. Petersons, P. Presetnick, J. Prugeer, G. Reiter, L. Rodrigues, W. Schorcht, M. Uhrin and V. Vintujlis. 2015. Return of the bats? A prototype indicator of trends in European bat populations in underground hibernacula. *Mammalian Biology* 80 (3): 170–7.

114 Rydell, J. J. Eklöf, H. Fransson and S. Lind. 2018. Long-term increase in hibernating bats in Swedish mines – effect of global warming? *Acta Chiropterologica* 20: 421–6.

115 Adams, R.A. and M.A. Hayes. 2008. Water availability and successful lactation by bats as related to climate change in arid regions of western North America. *Journal of Animal Ecology*. DOI: 10.1111/j.1365-2656.2008.01447.x

116 Moir, Monika, L.R. Richards, M. Cherry and R.V. Rambau. 2020. Demographic responses of forest-utilizing bats to pasts climate change in South Africa. *Biological Journal of the Linnean Society*. https://doi.org/10.1093/biolinnean/blaa048

117 Gilmour, L.R.V., M.W. Holderied, S.P.C. Pickering and G. Jones. 2021. Acoustic deterrents influence foraging activity, flight and echolocation behaviour of free-flying bats. *Journal of Experimental Biology* 224 (20). https://doi.org/10.1242/jeb.242715

Chapter 12

118 Eklöf, J. and J. Rydell. 2021. Attitudes towards bats in Swedish history. *Journal of Ethnobiology*. https://doi.org/10.2993/0278-0771-41.1.35

119 Rocha, R., A. López-Baucells and Á. Fernández-Llamazares. 2021. Ethnobiology of bats: exploring human–bat inter-relationships in a rapidly changing world. *Journal of Ethnobiology* 41: 3–17.

120 Stockman, L.J., L.M. Haynes, C. Miao, J.L. Harcourt, C.E. Rupprecht, T.G. Ksiazek, T.B. Hyde, A.M. Fry and L.J. Anderson. 2008. Coronavirus antibodies in bat biologists. *Emerging Infectious Diseases*. DOI: 10.3201/eid1406.070964

121 Fenton, M.B., S. Mubareka, S.M. Tsang, N.B. Simmons and D.J. Becker. 2020. COVID-19 and threats to bats. *FACETS*. DOI: 10.1139/faceets-2020-0028

[122] Fenton, M.B., A.C. Jackson, and P.A. Faure. 2020. Bat bites and rabies: the Canadian scene. *FACETS*. DOI: 10.1139/facets-2019-0066

[123] De Oliveira, M.B. and C.R. Bonvicino. 2020. Incidence of viruses in Neotropical bats. *Acta Chiropterologica*. DOI: 10.316115081109AC2020.22.2.018; Brook, C.E., and Dobson, A.P. 2015. Bats as 'special' reservoirs for emerging zoonotic pathogens. *Trends in Microbiology* 23(3): 172–80. PMID: 25572882 DOI: 10.1016/j.tim.2014.12.004

[124] Olival KJ, Hosseini PR, Zambrana-Torrelio C, Ross N, Bogich TL, and Daszak P. 2017. Host and viral traits predict zoonotic spillover from mammals. *Nature* 546(7660): 646–50. PMID: 28636590. DOI: 10.1038/nature22975; Mollentze, N. and D.G. Streiker. 2020. Viral zoonotic risk is homogenous among taxonomic orders of mammalian and avian reservoir hosts. *PNAS*. www.pnas.org/cgi.doi/10.1073/pnas.1919176117

Answer to the quiz on page 117

Two of the images show vampires. Clockwise from top left, these bats are: Common Big-eared Bat, Mesoamerican Moustached Bat, Common Vampire Bat, Common Vampire Bat, Yellow-throated Big-eared Bat, Proboscis Bat.

Index

References to photographs appear in *italic*

A

B

J

K

L

M

P

Q

R

S

T